Studies in Big Data

Volume 59

Series Editor

Janusz Kacprzyk, Polish Academy of Sciences, Warsaw, Poland

The series "Studies in Big Data" (SBD) publishes new developments and advances in the various areas of Big Data- quickly and with a high quality. The intent is to cover the theory, research, development, and applications of Big Data, as embedded in the fields of engineering, computer science, physics, economics and life sciences. The books of the series refer to the analysis and understanding of large, complex, and/or distributed data sets generated from recent digital sources coming from sensors or other physical instruments as well as simulations, crowd sourcing, social networks or other internet transactions, such as emails or video click streams and other. The series contains monographs, lecture notes and edited volumes in Big Data spanning the areas of computational intelligence including neural networks, evolutionary computation, soft computing, fuzzy systems, as well as artificial intelligence, data mining, modern statistics and Operations research, as well as self-organizing systems. Of particular value to both the contributors and the readership are the short publication timeframe and the world-wide distribution, which enable both wide and rapid dissemination of research output.

** Indexing: The books of this series are submitted to ISI Web of Science, DBLP, Ulrichs, MathSciNet, Current Mathematical Publications, Mathematical Reviews, Zentralblatt Math: MetaPress and Springerlink.

More information about this series at http://www.springer.com/series/11970

Marek Kretowski

Evolutionary Decision Trees in Large-Scale Data Mining

 Springer

Marek Kretowski
Faculty of Computer Science
Bialystok University of Technology
Bialystok, Poland

ISSN 2197-6503   ISSN 2197-6511   (electronic)
Studies in Big Data
ISBN 978-3-030-21853-9   ISBN 978-3-030-21851-5   (eBook)
https://doi.org/10.1007/978-3-030-21851-5

This Springer imprint is published by the registered company Springer Nature Switzerland AG
The registered company address is: Gewerbestrasse 11, 6330 Cham, Switzerland

Introduction

The world around us is changing very fast mainly because of spectacular progress in information and communication technologies. A phenomenon known only from science-fictional movies like a humanoid robot or an autonomous driving system becomes more and more natural and obvious for our children and maybe even sooner. Computerized systems seem indispensable in virtually every area of people's lives. Almost all activities can be registered, digitalized, and archived.

Besides traditional data sources like business and finance, a flood of other types of data with a more dynamic and complex nature has been observed. The Internet and especially social media are now an unstoppable volcano of unstructured texts, images, and video. Omnipresent sensors provide current and exact geographic-based data concerning weather, traffic, grids, and so forth. In the healthcare domain, clinical and imaging data can be complemented by huge and varied OMICS data (e.g., genomics, proteomics, or metabolomics). Without a doubt, we are overwhelmed by heterogeneous and vast data, and we should say that humanity has entered into the Big Data era.

With the widespread availability of enormous amounts of potentially useful data, a key question arises regarding if we are able to take advantage of such an opportunity. The traditional well-established machine learning and data mining methods, which worked well with nicely structured datasets composed of hundreds of instances and tens of features, cannot be directly applied to current volumes. Practitioners have started to understand that to try to discover interesting patterns and transform them into real knowledge, new approaches and tools are needed to preprocess and analyze even parts of these flooding data. We have to extend existing solutions or propose new ones to be able to cope with the millions of examples and thousands of features. And the first criterion for assessing the proposed methods is their actual applicability, which means a guarantee of receiving competitive results on the available resources in a reasonable time.

From a computational perspective, we have to remember that even with significant progress being made in the hardware, the computational power is obviously limited, and valid restrictions are still imposed on fast disks and operational memories. Simply speeding up the processor clock is not possible anymore, so research must focus on exploiting parallel and distributed processing solutions. Hopefully, in recent years, significant progress in the domain has been observed, especially with the advent of general programming on graphical processing units (GPGPU) and a distributed computing on commodity clusters based on the Hadoop ecosystem. These low-cost technologies create suitable conditions for developing new, efficient algorithms that are indispensable in the emerging data science discipline.

The second dimension, which should be taken into account when discussing predictive solutions, is the creditability of the applied techniques. The easiest way to convince a domain expert that the proposed method is trustworthy, is to enable him or her to understand the mechanisms of the algorithm and to explain why such a prediction is being proposed for a given instance. A lot of broadly defined artificial intelligence approaches, like neural networks, SVM, or ensembles, are very competitive in terms of their prediction accuracies, but their interpretability is low. They are often called black-boxes, which here is an insight into how the learning process and a rational justification of a specific decision are difficult. It is why more transparent methods are often preferred, especially, if the estimated prediction accuracy has not deteriorated. It allows the analysts to demystify the applied machine learning or data mining techniques and to concentrate on improving these techniques' performance.

Decision trees are one of the most popular white-box methods. A sequential analysis of identified conditions going from general to more specific is very close to the human way of reasoning. The resulting tree structure with tests in internal nodes and predictions in leaves can be nicely visualized. The tests, especially if they concern single feature, are easily interpretable. Each new prediction can be convincingly explained by tracing a path from the tree root to a leaf and then translating this path into a decision rule. Decision trees are traditionally induced with a greedy, top-down approach, which is relatively fast and effective for most practical applications. On the other hand, the induced trees are sometimes overgrown and not very stable. In this book, I will try to demonstrate that compact and accurate decision trees can be efficiently generated with the use of evolutionary algorithms, even for really large-scale data. This approach can be easily tailored for specialized applications.

My first contact with evolutionary computations was in the early 1990s during my computer science studies at Bialystok University of Technology, where I had a pleasure to participate in lectures from Prof. Franciszek Seredynski. They interested me so much that I decided to implement my first evolutionary algorithm and try to find possible applications. As a result, in one of my first publications, a genetic algorithm was applied as a tool for the optimal tolerance thresholds searching in tolerance rough sets. Then, for a few years, I worked with Wojciech Kwedlo on an evolutionary induction of decision rules.

My first publications on decision trees were devoted just to an oblique tree induction, but soon, I started to develop an evolutionary algorithm to search for oblique splits based on dipolar criteria. The algorithm was launched in each non-terminal node during the classical, top-down induction. Soon after, based on my experiences with the evolutionary generation of decision rules, I decided to abandon the top-down approach and started to investigate a global induction of decision trees. And finally, in 2004, we published our first report on the evolutionary induction of decision trees. Since then, we have been working on various tree types and applications. We put a lot of effort into showing that evolutionary induction can be a competitive alternative. After many years of work on evolutionary data mining systems, I really believe that the evolutionary approach is well suited for large-scale data mining, especially now with the advent of general computing on GPU.

In this book, I want to sum up my research conducted over the last 15 years on the evolutionary induction of decision trees. I hope that the book will facilitate the popularization of global induction and will help garner more interest in it.

Book Organization

The book is divided into four parts. In the first part, some basic elements from three domains are discussed, all of which are necessary to follow the proposed approach: evolutionary computations, decision trees, and parallel and distributed computing. These chapters should be treated as subjective excerpts from the vast and important disciplines. They do not pretend to be exhaustive; only the most important concepts and terms that are linked with the global induction of decision trees are presented. In the first chapter, the main steps of a generic evolutionary algorithm are provided, and the key decisions for designing a specialized algorithm are discussed. At the end of the chapter, the essential definitions and methods of multi-objective evolutionary optimization are also mentioned. The second chapter is devoted to data mining and decision tree fundamentals. The classical approaches to decision tree induction are briefly presented, and the related works on applications of evolutionary computation are collected. In the third chapter, distributed and shared memory programming are first referred to. Then, a general purpose programming on graphical processing units is discussed. Finally, Apache Spark, the most popular distributed framework for large-scale data, is introduced.

In the second part of the book, the proposed evolutionary approach to the induction of decision trees is presented in detail. In the first of two chapters, univariate trees for both classification and regression problems are discussed, whereas in the second chapter, oblique and mixed trees are shown. For all variants, suitable representations are proposed, and reliable and efficient initializations are introduced. Carefully developed specialized genetic operators that incorporate local search extensions ensure a good balance between exploration and exploitation during evolution. Moreover, various forms of single- and multi-objective fitness functions are investigated.

In the third part, two specific extensions of the generic approach are presented. They show that if an analyst has prior insights into the problem, this knowledge can be relatively easily incorporated into evolutionary induction, and it could result in more accurate and reliable predictions. As a first case study, a cost-sensitive prediction is investigated for both classification and regression problems. As a second example, a multi-test decision tree tailored to gene expression data is presented.

The last part of the book concentrates on the applicability of the proposed evolutionary approach for large-scale data. It is obvious that the strong use of parallel and/or distributed processing is indispensable for an efficient induction. Apart from the classical OpenMP/MPI techniques for computing clusters, more up-to-date solutions are investigated. It seems that a GPU-based approach fits in perfectly when it comes to scaling up an evolutionary induction. Finally, Apache Spark is adopted, and it pushes the size limits of mineable datasets.

The book concludes in the last chapter, and some directions of possible future works are sketched out.

Acknowledgements

First, I want to thank my former and current co-workers: Marek Grzes, Piotr Popczynski, Krzysztof Jurczuk, Marcin Czajkowski, and Daniel Reska. I had the chance to serve as a supervisor for their Ph.D. and M.Sc. theses. This book is based on the results of my fruitful collaboration with this group of very talented and hardworking scientists. I am sure that they will succeed in science or industry, and I hope that we will collaborate in the future.

I would like to express my gratitude to my teacher, Prof. Leon Bobrowski. He offered me a position on his team when I was starting my academic career. Then, he sent me to the University of Rennes 1 for preparing a jointly supervised Ph.D. thesis, where I met excellent people and learned a lot. I had the pleasure to be his deputy when he served as the dean at our faculty, and when he decided to retire after more than 25 years, he offered me the lead position.

I want to thank my friends at the Faculty of Computer Science: Agnieszka Druzdzel, Wojciech Kwedlo, and Cezary Boldak for all their help and support.

Finally, I want to deeply thank my whole family, especially my wife, Malgorzata, and children, Jacek and Izabela.

Contents

Part I
Background

Chapter 1
Evolutionary Computation

A lot of typical problems that have to be commonly solved in engineering or business can be formulated as optimization problems. The performance of an activity or the value of a decision are characterized by a certain cost function, and here, possible alternatives are considered. The goal is to find the best solution from all feasible solutions because some constraints can be also imposed. There exist dozens of classical methods to solve various types of optimization problems (continuous or discrete, constrained or unconstrained, etc.). Typically, partial solutions are combined to obtain the final solution, or the performance of a trial solution is improved iteratively based on the adopted strategy. The choice of the next step can be deterministic (e.g., based on the gradient) or stochastic; it can also rely on previous or historic steps. These techniques usually provide fast converge to locally optimal solutions, but for many problems (e.g., highly multimodal), this is not enough. It is especially evident for so-called hard optimization problems, where a search space could be huge, nonlinear, and discontinuous.

Considering only one trial solution seems to be a real limitation of the traditional methods. An approach that simultaneously investigates more candidate solutions can be treated as an imposing alternative because it can very easily extend the search area. Moreover, it also offers the possibility of providing interactions between these candidate solutions. In recent decades, many meta-heuristic methods have been proposed [1], and these methods process a group of candidate solutions. A lot of them are inspired by some biological or physical processes. Among them, evolutionary computation [2] is one of the most successful groups of algorithms. Typically, the representatives of this group are called evolutionary algorithms. There are many variants, but they share main ideas because they all were inspired by Darwin's principles of evolution and natural selection.

Evolutionary algorithms [3] are iterative techniques where a population of individuals (called chromosomes) representing candidate solutions evolves. Individuals are capable of reproduction and are subject to genetic variations. The environmental pressure determines the quality of the solutions (expressed as a value of the fit-

© Springer Nature Switzerland AG 2019

M. Kretowski, *Evolutionary Decision Trees in Large-Scale Data Mining*,
Studies in Big Data 59, https://doi.org/10.1007/978-3-030-21851-5_1

ness function), and better-adapted individuals a better chance of surviving and being present in the next generation. New chromosomes can be created through crossover, where two (or more) individuals (called parents) are combined to create at least one offspring (child). A mutation introduces new traits in individuals to increase diversity. After applying genetic operators to a population, the resulting individuals are evaluated according to the fitness and solutions that will be maintained in the iteration; these will be selected or reproduced according to the adopted selection strategy. The simulated evolution is finished typically after a predefined number of generations or more complex stopping rules have been developed. It should be clearly stated that evolutionary algorithms are just inspired by the biological evolution and do not try to exactly mimic nature.

There are many distinct origins of the current version of evolutionary computation. The following are the most recognized and influential approaches [4]:

- Genetic algorithms [5]—typically operate on fixed-size chromosomes with binary coding. This representation prefers standard genetic operators. A crossover, which exchanges genetic material between two individuals, is a primary operator and is applied with a high probability. A mutation that creates a slight perturbation with a low probability is perceived as a secondary operator.
- Evolutionary strategies [6]—use problem-dependent representations, such as a vector of real values for continuous function optimization. A mutation is the main operator, and its ranges are self-adapting. The relations between a population of parents and a population of modified individuals define various types of strategies based on the population sizes and selection rules. Mutants can compete with parents or can be disregarded.
- Genetic programming [7]—was proposed as an automated method for evolving computer procedures. The program is typically represented as a syntax tree of a dynamically changing size. The variables and constants of the program (called terminals) are leaves, whereas the internal nodes represent (arithmetic) the operations (called functions). The terminals and functions define the alphabet of the program. Many variants of specialized differentiation operators can be proposed, with a subtree exchange between two trees and a subtree mutation as the most basic ones.
- Evolutionary programming [8]—with a representation typically tailored to the problem domain (originally, individuals represented as finite state machines). Similar to evolutionary strategies for real-valued vector optimization, a string of real values is evolved. New individuals are created only as a result of a mutation, which typically adds a random number from a given distribution to a parent. Any kind of recombination is applied in the conventional approach, but later, sexual reproduction is also introduced.

Application-specific variants of the evolution-based procedures have also been proposed. Learning classifier systems [9] are a good example of an approach developed for rule-based machine learning. They try to combine a discovery component that is based on genetic algorithms with a (supervised or unsupervised) learning component, and they can be perceived as an encouraging motivation of an evolutionary data mining approach presented in this book. As for the continuous global optimization problems, a differential evolution [10] is currently the most popular and successful evolutionary algorithm. In this family of methods, the mutation is typically performed by calculating the vector differences between other randomly selected individuals and adding it to a randomly selected base vector. In the case of a crossover, a mutant vector is combined with components of the corresponding population vector. The population individuals compete with the created trial individuals to determine which ones will be maintained in the next generation.

Initially, various paradigms have been developed independently, but over time, more and more inevitable interactions between de facto similar approaches have been observed. Successful ideas from one method were adopted and integrated into other methods. New hybrid algorithms were proposed, yet assigning them to classical paradigms has become problematic and not really useful. Such a noticeable crossbreeding of ideas led to the formation and universal acceptance of evolutionary computation as a unified scientific discipline [11]. All the methods developed under this wide umbrella are called evolutionary algorithms. So now, evolutionary algorithms are mainly designed to solve hard optimization problems and search complex spaces, and they are capable of adapting to changing environments.

It is also worth mentioning that these evolutionary methods can be beneficially combined with traditional local search techniques within so-called memetic algorithms [12]. Typically, individuals can be locally improved in a separate operation in every generation, or a local optimization is embedded into a mutation. In certain conditions, these extensions can substantially speed up the search while preserving the robustness of the approach, leading to numerous, important, and practical applications [13].

1.1 The Algorithm

The standard iterative process of the evolutionary algorithm is schematically presented in Fig. 1.1. The specific realizations can be slightly different, but the main components and activities are almost invariable.

The process starts with a population initialization. Typically, an individual could be randomly generated, but it should be ensured that the resulting initial population will be highly diverse and encompass the whole search space because this obviously facilitates the evolution. Then, these new individuals can be assessed by means of a fitness function, and the main loop of the algorithm can start. All iterations are the same: individuals are selected for a reproduction, and differentiation operators are applied, the modified or created individuals are evaluated, and a succession is

Fig. 1.1 Typical evolutionary (genetic) algorithm process

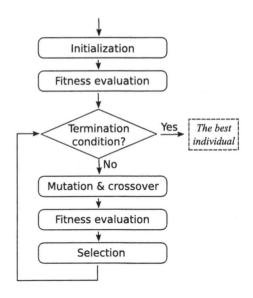

performed when the next population is formed. After each generation, a termination condition is verified. In the simplest scenario, the fixed number of iterations is simulated, or time constraints are imposed. With more advanced solutions, certain properties of the population devoted to the algorithm convergence can be observed, and the algorithm is stopped when the expected pattern has been reached. An individual with the best-found fitness[1] is returned as the final result of the algorithm.

In each generation, the individuals are differentiated by applying genetic operators. There are two typical operators: mutation, which modifies a single individual, and crossover, which usually involves more individuals and generates at least one offspring. An individual differentiation is the primary mechanism in the evolutionary search and should guarantee that a proper balance between exploration (checking new, often distant locations) and exploitation (detailed searching near the current position) is preserved. The mutation randomly perturbs an individual chromosome in a fragmentary way, and more frequently, this results in a small displacement of the original individual in the search space. More distant modifications are obviously also possible if the alterations reach the most significant positions in the chromosomes. Concerning the crossover, where the parents are usually fully recombined, the offspring are more different from the originals because the number of modifications is naturally greater compared with a local mutation. This results in distant transpositions, which could be especially fruitful when an exploration of new areas is necessary.

Selection mechanisms are the key driving forces of this type of simulated evolution. Individuals that are better adapted to environmental conditions expressed in terms of the fitness should have a better chance of surviving than less-fitted indi-

---

[1]The best-fitted individual is not typically detected in the last generation.

viduals. Only reproduced and selected individuals will take part in the creation of the descendants. This mechanism greatly influences the search directions and could be decisive when it comes to the efficiency and effectiveness of the evolution. A good compromise between providing a sufficient diversification of a population and promoting the most promising individuals must be established. Too much focus on the best individuals can result in a premature convergence on only locally optimal solutions. On the other hand, not enough support for the better individuals will substantially slow down the search and will waste computational resources. Clearly, maintaining the right balance in different phases of the algorithm is not the easiest task, but the selection algorithms developed over the years make achieving this goal possible [14].

When one wants to apply the evolutionary approach to a given problem, it is usually profitable to go beyond the standard and try to develop a specialized version of an evolutionary algorithm. The best results are obtained if the knowledge of the problem specificity can be incorporated into the algorithm [4] and can be used to select the most adequate mechanisms from the wide range of available elements. In many applications, this allows for a significant reduction in the search space, and as a result, quickly finding better solutions. As can be expected, in such an individualized approach, it is necessary to choose the values of many regular (from the standard algorithm), but also additional, control parameters (concerning the specialized elements). Evolutionary algorithms are known to be relatively insensitive to minor changes in basic parameter values [15], but typically, a series of experiments with varying values is performed to find the best settings. This tuning can be done manually or automatically, which can be expected to be more efficient, but it is obviously much more computationally demanding. Different ways of parameter control in evolutionary algorithms are discussed in [16].

When a new, specialized evolutionary algorithm is being designed, there are a few main decisions that must be made [17]:

- How a solution will be represented;
- how an individual will be initialized;
- how the proper fitness function should be defined;
- how effective genetic operators should be developed, ones that are coherent with the adopted representation; and
- how the convenient selection mechanism should be chosen.

It is obvious that these components of the algorithm are highly interdependent and could not be considered in isolation from the other elements. For example, genetic operators should be aware of the representation constraints to avoid unacceptable or degenerated individuals. Similarly, a selection mechanism should be aware of the fitness evaluation because it can rely on a range of fitness values. The definition of the fitness function should be as consistent as possible with the goal of an attacked problem. However, in certain situations, an exact mapping is not possible, like, for example, in prediction tasks, where we are interested in optimizing an accuracy, so here not on the training, but on the unseen data. Algorithm designers should try to ensure that the global optimum of the fitness function coincides with (or at least is very close to) the best actual solution. In addition, the fitness evaluation of any individuals should be consistent and should promote better solutions in the whole search space.

1.2 Problem Representation

In evolutionary computation, numerous inspirations from molecular biology are present. One of them is a distinction between a genotype, which defines a structure of evolved genetic information, and a phenotype, which characterizes the actual features and/or behaviors assigned to an individual. In certain evolutionary algorithms (e.g., in genetic algorithms), there exists a mapping between a solution encoding (an individual representation) and a solution itself. In other approaches, this distinction is not necessary because individuals are not specially encoded, and they are directly processed. Anyway, an individual representation should reflect all the features that are relevant to the evaluation of its quality.

In a standard genetic algorithm [18], a fixed-size binary chromosome is used to represent an individual. This means that any feature that is not binary has to be encoded as a binary substring. In the case of a continuous-valued feature, a representation precision can be easily increased by just using more bits. With this type of representation, the standard genetic operators can be typically utilized, which simplifies a problem-specific application and can be seen as an important advantage for the meta-heuristic algorithm. Unfortunately, a natural binary coding can have some disadvantages. For example, the impact of a local change (flipping one bit) on the genotype level can have a very diverse effect on the phenotype level depending on the position of the bit (a modification of the most significant bit of a number versus the less significant bit). As for a nominal variable, the binary representation is lengthy, which can create superfluous computational costs. And when more complex coding is imposed, redundancies and inconsistencies can be easily introduced as a result of a simple differentiation. In Fig. 1.2a, an example of the binary genotype and the corresponding phenotype is presented.

In many problems where the numeric variables are processed, a floating-point representation of a single variable seems to be much more natural. For continuous optimization, the fixed-size vectors of floating point numbers are typically evolved, and genotype-phenotype mapping can be abandoned. Interestingly, this type of straightforward representation can be extended by introducing additional elements that have evolved at the same time; this mechanism can be dedicated to a self-adaptation. In evolutionary strategies, an additional vector of the same size is appended, and it stores standard deviations corresponding to the processed features. These evolving standard deviations are used for controlling the mutation ranges. In Fig. 1.2b, an example of an extended real-number representation for a self-adaptation is presented.

The representations mentioned so far have been of a fixed length, but for certain problems, a variable-length chromosome is indispensable, and the proper size of an individual is optimized as well. Moreover, if a non-standard representation is intro-

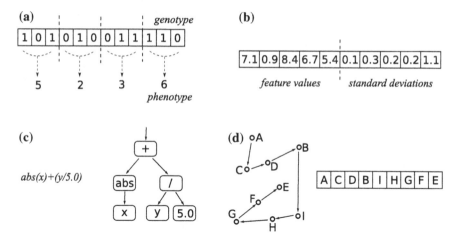

Fig. 1.2 Various representation examples: **a** a binary representation with a genotype-phenotype mapping, **b** evolutionary strategies, **c** a symbolic expression with the corresponding tree representation, **d** a path representation in a TSP problem

duced, it inevitably entails the need to develop specialized differentiation operators. One of the most popular variable-length representations is a tree-based representation from genetic programming (Fig. 1.2c).

A traveling salesman problem (TSP) is a classical search problem where the shortest route from a given town that visits all towns and returns to the starting place is sought. The towns can be numbered, so a permutation of these numbers can be treated as a natural representation of the searched route. It is a so-called path representation, but more sophisticated propositions, like the adjacency representation, have been developed [19]. In Fig. 1.2d, an example of the path representation for the TSP problem is depicted.

1.3 Genetic Operators

A good representation is a necessary condition for an efficient evolutionary algorithm because it can substantially limit the search space. The next issue, which is even more crucial for a smooth and robust convergence, is a stochastic mechanism for generating candidate solutions based on the existing ones. Any region of the search space can be explored if at least one individual located in the area will appear as a result of a differentiation (or initialization) and can be evaluated. Conventionally, two types of differentiation operators are distinguished depending on the number of individuals involved. Here, the term mutation is typically used when a single chromosome is randomly modified. Typically, a mutation has only a limited impact on a genotype, but the influence of a local change on a phenotypic interpretation can be significant.

In other words, a distance in the solution space between the original individual and the mutated one can be very diverse. This clearly reveals that a mutation can play an important role in both exploration and exploitation.

The operators in the second group concern more than one individual, and usually, the term crossover is utilized, when two parents exchange genetic information and offspring are generated; sometimes, the term recombination is preferred when more individuals are involved. The process is obviously inspired by a natural reproduction, and the offspring should derive some features from their parents. The rationale is that a combination of the features from already existing individuals can potentially lead to an even better-adapted descendant. Concerning the distances between the parents and offspring, they are typically larger than in a mutation because the larger parts of the chromosomes are modified. On the other hand, the mechanism enables a fast dissemination of well-solved subproblems.

Sometimes, it is not so easy to decide to which category a genetic operator should be assigned as with the differential evolution operator.

It is worth noting that in the case of a specialized representation, there is a strong relationship between a chromosome structure and the genetic operators. Only a good understanding of the representation allows a designer to propose adequate and efficient operators that can find the right balance between exploration and exploitation. Moreover, it should not be forgotten that an operator cannot deterministically ensure better fitness because a stochastic character of differentiation is a key factor in the robustness of an evolutionary algorithm.

1.3.1   Mutation

The purpose of a mutation operator is the introduction of random modifications in an individual's chromosome, which enables the preservation of a diversity of a population and prevents of premature convergence. It can also be seen as a fundamental mechanism for inspecting the search space. Typically, any gene from a chromosome can be mutated independently with a relatively low probability. As a rule, a result of the mutation should not be too strong because it can completely destroy the fitness that has been achieved to this point. Moreover, it is usually expected that small perturbations will be more likely than large modifications.

In a standard genetic algorithm with a binary representation, a mutation of a single position is realized through bit flipping. By adjusting a mutation probability (by taking into account a chromosome length), it can be easily obtained that on average only one position (one bit) is modified. It sounds reasonable, but it should be remembered that if binary coding is applied to represent, for example, integers, the impact from a different bit mutation would be very diverse depending on the significance of the position.

When a single position is represented by an actual number (e.g., integer or double), a mutation is typically realized by a random perturbation of the current value. The problem with the various significances of chromosome positions disappears because

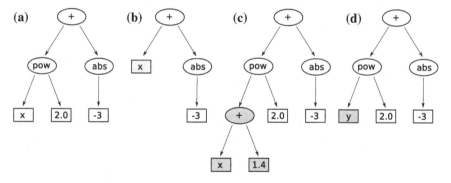

Fig. 1.3 Examples of a tree mutation: **a** original tree, **b** pruned one, **c** with an additional subtree, **d** with a modified symbol

there is only one gene for a feature. However, another problem with the determination and control that occur during the evolution of a mutation range now comes into play. A default range for a float number is huge, and an adoption of a reasonable deviation scope for every feature could be difficult. It is obvious that well-chosen ranges are necessary for a proper convergence of the algorithm.

In evolutionary strategies, the problem is mitigated by introducing self-adaptation mechanisms. For every feature, a current mutation strength (step size) is stored and can be modified according to the achieved results of the mutation. One of the most well-known adaptive control of step sizes is the 1/5-th success rule by Rechenberg [20], which increases the step size if the success rate is over 1/5-th and decreases it if the success rate is lower. In a differential evolution, a perturbation is related to a difference between two individuals, which can be seen as another smart mechanism of counteracting the problem.

In genetic programming using a tree-based representation, a node is randomly selected and modified. Because a tree is a hierarchical structure, the importance of the modification is highly dependent on a position of the selected node in the tree. Obviously, a mutation of the tree root will affect the whole tree, whereas the modification of a leaf will have only a very local meaning. Moreover, various types of modifications should be distinguished. These modifications can refer mainly to a tree structure when an internal node is pruned into a leaf with a terminal symbol or when a terminal node is replaced with a subtree. They can also concentrate only on the substitution of a symbol associated with a node without intensional change of the tree size. In Fig. 1.3, mutation examples are presented.

As for the path representation in the TSP problem, a mutation cannot encompass only one position cause it then will violate the phenotypic interpretation. The simplest route modification involves changing the order of the two visiting locations; hence, two positions in a chromosome need to be modified in a coordinated manner.

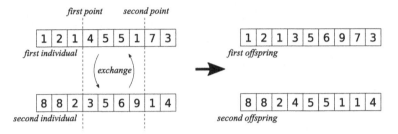

Fig. 1.4 Two-point crossover example

1.3.2 Crossover

The role and importance of a recombination as a search operator was different in the initial evolutionary approaches but now is commonly accepted and even theoretically proved [21]. Apart from a reproduction, it is a mechanism that clearly distinguishes an evolving population in the evolutionary algorithm from a simultaneous evolution of separate individuals. It should be noted that especially in the case of specialized representations, an elaboration of suitable crossover operations can be a challenging task, mainly because of the need to comply with the limitations of the representation and the expectation of the similarity of the descendants to the parents.

The most typical form of genetic information exchange between two linear chromosomes of the same size is a one-point crossover. A position of the chromosome split is randomly chosen, and substrings starting from the position are exchanged, generating two new offspring. These new offspring can be easily extended into a two-point crossover when the two split points are introduced and when only the substrings between these two points are exchanged. Usually, if the genotype-phenotype mapping is presented, the split positions should be cautiously chosen so as not to destroy the phenotypic features. An example of a two-point crossover is presented in Fig. 1.4.

A feature value that appears in the offspring does not need to be a copy of one parent feature value, but it can be also a mixture of both parent values. For example, in an averaging crossover, every gene value of the offspring lies between the corresponding values of the parents. Typically, a difference vector is calculated between two parents, and its random fraction[2] is added to one parent. The second offspring (if it exists) is symmetrical with respect to the center of the line segment joining the parents.

If a non-linear representation is approved, like the tree-based representation in genetic programming, an exchange of genetic information can be slightly more complicated. In the simplest scenario, two nodes are randomly chosen in the parents, and if there are no syntactic contraindications, the corresponding subtrees are replaced, and two offspring are generated. However, it should be noted that in genetic pro-

---

[2]A fraction coefficient between zero and one is randomly chosen from the uniform distribution.

gramming because of, among others, the bloat and intron problems,[3] the crossover is often perceived as being too destructive [22].

Even for the fixed-size linear chromosomes composed of natural numbers, it is sometimes not possible to apply straightforward crossover solutions. In the case of path representation, the standard one-point crossover will result in unacceptable offspring corresponding to the routes with repeated or missing towns. Smarter procedures are needed [23], and a partially-mapped crossover is a good example. This crossover builds offspring by choosing only a subsequence of a tour from one parent and preserving the order and position of as many towns as possible from the other parent. More precisely, two subsequences are randomly chosen by selecting two cut-off points serving as boundaries for the swapping operations because two offspring are generated. Similarly, in an ordered crossover, two subsequences are also randomly chosen, but then, an attempt at preserving the relative order of the towns from the other parent is made.

1.4 Reproduction/Selection

In most evolutionary algorithms, the population size does not change between generations and is rather modest (less than 100 individuals). However, during one iteration, the number of considered individuals can be higher, but surplus individuals are not included in a new population. Various mechanisms for population selection [11] are applied, and they can be deterministic or stochastic. New individuals are created as the results of genetic operator applications, and they can substitute for the parents or can compete with them. Moreover, additional copies of the well-fitted individuals can be generated, which obviously leads to a reduction of a space in the population for the less-fitted ones. From an efficiency perspective, we are typically interested in preserving the best-found individual (or individuals), and this type of approach is called an elitist strategy. It is mainly useful for stochastic selections because for most deterministic methods, the best individuals are not omitted. But one should be careful when increasing the selective pressure [3] because the diversity of the population can be decreased too quickly, potentially leading to a premature convergence in a local optimum.

Below, the most popular stochastic selection methods are briefly discussed.

---

[3]In genetic programming, bloat can be described as excessive code growth within the individuals of the evolving population without a proportional improvement in fitness, whereas introns are (redundant or unproductive) the parts of the code that do not contribute to the calculation that is being made.

1.4.1 Proportional Selection

A proportional selection is one of the most straightforward and simple methods, and its popularity is increased by involving a roulette wheel mechanism, which could sound attractive to newcomers. Because any selection mechanism should prefer better individuals, in the proportional selection, this preference is directly related to the differences in the fitness function values. Individuals are repeatedly and randomly chosen (with return) from a selection pool. In each draw, the probability $P(I_i)$ that an individual I_i is selected depends on its absolute fitness value $F(I_i)$ compared with the sum of the absolute fitness values of the rest of the population:

$$P(I_i) = \frac{F(I_i)}{\sum_{j=1}^{L} F(I_j)}, \tag{1.1}$$

where L is both a selection pool size (typically equal to a population size) and a number of selected individuals. The sum of all probabilities $P(I_i)$ equals 1.0.

A roulette wheel method can be applied to perform a draw according to the probabilities assigned to all the individuals from a selection pool. First, a sequence of numbers $t_0 = 0, t_1, \ldots, t_L = 1$ is defined in such a way that:

$$t_i = \sum_{j=1}^{i} P(I_j). \tag{1.2}$$

In each draw, a number r is randomly chosen from a uniform distribution over the range $(0, 1]$, and an individual I_i is selected for which the condition $t_{i-1} < r \leq t_i$ holds. Conceptually, it is the same as spinning a one-armed roulette wheel where the sizes of the holes reflect the selection probabilities.

It should be noted that relying on the absolute fitness values can problematic in certain conditions [2]. It is possible that the fitness values will differ significantly between the individuals (by an order of magnitude, for example). In this type of situation, outstanding individuals take over the entire population very quickly, preventing a profound exploration. This often leads to what is called premature convergence, where overlooking better existing solutions is probable. On the other hand, when all the fitness values are very close to each other, there is almost no selection pressure. A selection becomes almost uniformly random, and an evolution converges very slowly, which may be unacceptable for many applications. To mitigate these aforementioned problems, various techniques can be introduced, such as windowing or (linear) scaling, but the most prominent idea is based on replacing an absolute fitness value by a fitness ranking position.

1.4.2 Ranking Selection

In a rank-based (ranking) selection, a constant selection pressure can be preserved. Individuals are first sorted according to the fitness function values. Next, an adjusted fitness $F_R(I_i)$ for each individual I_i can be calculated based on its position $Rank(I_i) \in 1, \ldots, L$ in the created ranking and by the fixed value of a selective pressure $\xi \in [1, 2]$. It is usually assumed that on average, an individual of median fitness should have one chance to be reproduced, which explains why the pressure needs to be limited. In the case of the most typical linear mapping, the fitness is equal to:

$$F_R(I_i) = (2 - \xi) + 2(\xi - 1) \cdot \frac{Rank(I_i) - 1}{L - 1}. \tag{1.3}$$

In a generational selection, ξ can be interpreted as the expected number of offspring (selections) allotted to the fittest individual [2]. Based on the adjusted values of the fitness function, the corresponding probabilities are calculated in the same way as in a proportional selection. However, these probabilities are related only to ranks, so they are constant through all generations for the individuals at the given position in rankings. In addition, in this case, non-linear mappings can be also applied, for example, exponential ones.

1.4.3 Tournament Selection

A completely different idea compared with proportional or ranking selections is a tournament selection. In this method, a calculation of complicated selection probabilities can be avoided. For selecting L individuals, L tournaments need to be carried out, and every tournament is organized with the same number of competitors $k, k \in \{2, \ldots, L\}$. Binary tournaments ($k = 2$) are the most popular. Individuals that participate in any tournament are randomly chosen using a uniform probability distribution from the whole selection pool (population). In each tournament, the best individual according to a fitness function is announced. Here, one individual can be a winner in a few competitions and will be selected to a new population repeatedly.

Interestingly, the implicit distribution of binary tournaments can be shown to be equivalent (in expectation) to a linear ranking, and when a tournament size is increased ($k = 3$), the implicit probability distribution changes to a quadratic ranking with more probability mass shifted toward the best individual [11]. The tournament selection becomes more selective with each increase in the tournament size.

1.5   Evolutionary Multi-objective Optimization

So far in this chapter, we have assumed that a fitness function reveals a single objective of a given optimization or search problem. However, this typical scenario is not always sufficient. Sometimes, we have two (or even more) objectives that are somehow conflicting or not simply convergent. One of the simplest approaches for overcoming this problem is introducing a unified fitness function that combines partial objectives and/or imposes mutual relations among them. This can be realized by the objective weighting or setting of a hierarchy (or priorities) of objectives. In many situations, it is complicated and difficult to carry out, for example, when one tries to unify completely different physical quantities. Here, when one a priori defines artificial weights or proposes ad hoc threshold values, the actual effects of these decisions are rarely understood. It seems that in many practical situations, it could be better to see and study a few various trade-off variants and only then to more reasonably choose the best solution after taking into account the real preferences of the user.

Because of their flexibility, evolutionary algorithms could be successfully applied to transform multi-objective problems into single-objective ones; also, as population processing methods, they are really well suited for a direct multi-objective optimization. Evolutionary multi-objective optimization (EMO) [24] has been a very active research area in recent decades, and many significant results have been achieved. Later in this chapter, the theoretical foundations and the most-known techniques used in EMO will be briefly presented.

1.5.1   Multi-objective Optimization Fundamentals

Let us consider Z potentially conflicting objectives f_i ($i \in \{1, \ldots, Z\}$) that need to be minimized simultaneously, where $f_i(a)$ denotes a value of the i-th objective of a given solution a. A vector $f(a) = [f_1(a), \ldots, f_Z(a)]$ represents the solution a in a multi-objective space. Two vectors (each representing a solution) are said to be incomparable when for each vector there exists at least one objective for which its value is superior compared with the other vector. A solution a is said to dominate (in Pareto sense) a solution b, which can be denoted as $a \prec b$, if $f_i(a)$ is not worse than $f_i(b)$ for any objective i and if $f_j(a)$ is better than $f_j(b)$ for at least one objective j:

$$a \prec b \Leftrightarrow \forall_{i \in \{1,\ldots,Z\}} f_i(a) \leq f_i(b) \wedge \exists_{j \in \{1,\ldots,Z\}} f_j(a) < f_j(b). \qquad (1.4)$$

Non-dominated solutions are called Pareto optimal, and the Pareto optimal set contains only non-dominated solutions:

$$\{a : \neg(\exists b : b \prec a)\}. \qquad (1.5)$$

Fig. 1.5 An example of the
Pareto front for a
two-objective problem

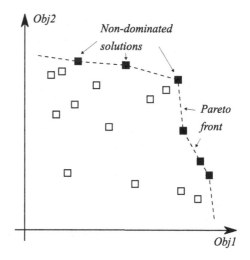

The set of all Pareto optimal solutions is referred to as the Pareto front. So the goal of the multi-objective optimization is to find out the Pareto front, which is composed of multiple alternative solutions, and present them for the decision makers. An example of the Pareto front for a two-objective problem is presented in Fig. 1.5.

Many multi-objective evolutionary algorithms that search for a set of Pareto-optimal solutions in a single run of the algorithm have been proposed so far [25]. Most show a fast convergence to a sub-set of solutions from the Pareto-optimal front and enough diversity of solutions to represent the entire range of the front. Among the most popular approaches are the Strength Pareto Evolutionary Algorithm (SPEA) [26] and Non-dominated Sorting Genetic Algorithm (NSGA-II) [27].

1.5.2 NSGA-II

The NSGA-II procedure focuses on emphasizing non-dominated solutions, and it relies on the elitism principle. Moreover, an explicit diversity-preserving mechanism is built into the selection process.

At each generation, based on the current population, a new offspring population (of the same size) is created in a standard way by applying genetic operators. Then, these two populations are combined, and a non-dominated sorting is launched. Individuals are ordered according to non-dominance classes (fronts). In the first class, only completely non-dominated solutions are included. In the next step, all individuals from the first class are disabled, and the second class of non-dominated individuals is determined. The process is repeated until all individuals are assigned to classes. Based on the non-dominance classes, individuals are gradually included into a new population starting from the first class. Because only half of the existing individuals can be preserved, only a limited number of fronts can be fully included. Concerning

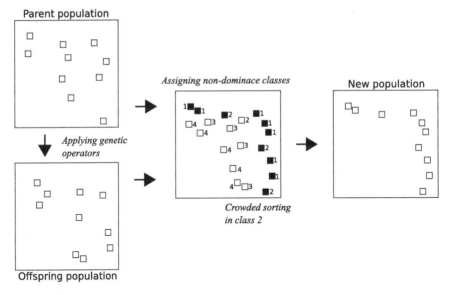

Fig. 1.6 A scheme of the NSGA-II procedure

the last front, which can be partially preserved, a special crowded sorting is used to choose the needed number of individuals. The crowding distance of an individual is a measure of the objective space around this individual not occupied by any other solution in the population.[4] The individuals are arranged in descending order of their crowding distance values, and top-ordered ones are the last individuals kept. All remaining individuals are discarded. A NSGA-II procedure scheme is depicted in Fig. 1.6.

The original NSGA-II, which is the leading approach, has been extensively investigated, and several extensions have been proposed. An Efficient Non-dominated Search (ENS) [28] is one of the more recent improvements of a search strategy, which reduces the complexity, especially in the case of a small number of objectives. In typical non-dominated sorting approaches, each solution has to be compared with all other solutions before assigning it to a front. An ENS first sorts a population according to one objective. Then, solutions are processed one by one to decide whether to include the current one in the front, but only comparisons with solutions that have already been assigned to the front are performed. In this way, a lot of redundant dominance comparisons can be avoided, and the computational efficiency is clearly improved.

Another improvement upon the original NSGA-II procedure deals with the crowding distance calculation. When two or more individuals share the same values for every objective, the associated crowding distances can be distorted and instable. This

---

[4]A crowding distance can be estimated by the perimeter of a cuboid formed by using the nearest neighbors in the objective space as the vertices.

problem could impact negatively the parameter space diversity preservation mechanism. Fortin and Parizeau [29] proposed a correction where the crowding distances are computed on unique fitnesses instead of individuals, improving both the convergence and diversity of NSGA-II. In this method, for all identically characterized solutions, the same crowding distance is assigned. Moreover, a special treatment for these solutions based on a round-robin technique is embedded into the last front selection. The diversity in the objective space is enforced by requiring individuals with different fitnesses to be selected before selecting higher crowding-distance points again.

References

1. Boussaid I, Lepagnot J, Siarry P (2013) Inf Sci 237:82–117
2. Eiben A, Smith J (2015) Introduction to evolutionary computing, 2nd edn. Springer, Berlin
3. Back T (1996) Evolutionary algorithms in theory and practice: evolution strategies, evolutionary programming, genetic algorithms. Oxford University Press, Oxford
4. Michalewicz Z (1996) Genetic algorithms + data structures = evolution programs, 3rd edn. Springer, Berlin
5. Holland J (1975) Adaptation in natural and artificial systems. University of Michigan Press, Ann Arbor
6. Schwefel H (1981) Numerical optimization of computer models. Wiley, Hoboken
7. Koza J (1992) Genetic programming: on the programming of computers by means of natural selection. MIT Press, Cambridge
8. Fogel L, Owens A, Walsh M (1966) Artificial intelligence through simulated evolution. John Wiley, Hoboken
9. Holland J (1980) Int J Policy Anal Inf Syst 4(3):245–268
10. Storn R, Price K (1997) J Glob Optim 11(4):341–359
11. De Jong K (2006) Evolutionary computation: a unified approach. MIT Press, Cambridge
12. Moscato P (1989) On evolution, search, optimization, genetic algorithms and martial arts: toward memetic algorithms. Caltech concurrent computation program, 790
13. Krasnogor N, Smith J (2005) IEEE Trans Evol Comput 9(5):474–488
14. Blickle T, Thiele L (1996) Evol Comput 4(4):361–394
15. Sipper M, Fu W, Ahuja K, Moore J (2018) BioData Min 11:2
16. Eiben A, Hinterding R, Michalewicz Z (1999) IEEE Trans Evol Comput 3(2):124–141
17. Kretowski M (2008) Obliczenia ewolucyjne w eksploracji danych. Globalna indukcja drzew decyzyjnych, Wydawnictwo Politechniki Bialostockiej
18. Golberg D (1989) Genetic algorithms in search, optimization, and machine learning. Addison-Wesley, Boston
19. Larranaga P, Kuijpers C, Murga R, Inza I, Dizdarevic S (1999) Artif Intell Rev 13(2):129–170
20. Schwefel H (1995) Evolution and optimum seeking. Wiley, Hoboken
21. Doerr B, Happ E, Klein C (2012) Theoretical Computer Science 425:17–33
22. Freitas A (2002) Data mining and knowledge discovery with evolutionary algorithms. Springer, Berlin Heidelberg
23. Michalewicz Z, Fogel D (2004) How to Solve It: Modern Heuristics, 2nd edn. Springer
24. Deb K (2001) Multi-Objective Optimization Using Evolutionary Algorithms. John Wiley & Sons
25. Zhou A, Qu B, Li H, Zhao S, Suganthan P, Zhang Q (2011) Swarm and Evolutionary Computation 1(1):32–49
26. Zitzler E, Thiele L (1999) IEEE Trans Evol Comput 3(4):257–271

27. Deb K, Pratap A, Agarwal S, Meyarivan T (2002) IEEE Trans Evol Comput 6(2):182–197
28. Zhang X, Tian Y, Cheng R, Jin Y (2015) IEEE Trans Evol Comput 19(2):201–213
29. Fortin F, Parizeau M (2013) Revisiting the NSGA-II crowding-distance computation. In: Proc. of GECCO'13, ACM pp 623–630

Chapter 2
Decision Trees in Data Mining

In this chapter, I explain what happened to make data become so much more available and where Big Data emerged from. I will show what can be searched for in these data and what tools are needed for mining the data. The differences and similarities between a classification and regression are described. Then, the focus is moved to decision trees and classical methods in their induction, but the presentation should not be treated as an extensive overview of this wide area of research. The most important information about decision trees is provided, and this subjective selection is intended to be helpful in understanding the proposed global approach. Finally, the related works on applying evolutionary computation in decision trees are studied.

2.1 Big Data Era

Over the last twenty years, there have been significant changes in the perception of data as sources of potential knowledge. At the end of the last century, the possibilities of the mass collection and storage of information about phenomena, activities, and processes were still severely limited. Interesting data were rare and usually difficult to obtain. The technological peak of the processing of structured data were relational databases or data warehouses. Data carriers in the form of traditional hard drives were expensive and unreliable, so there were no ways to procure straightforward data integration.

Fortunately, the situation started to change with the emergence of several interlinked ICT processes. Above all, the spread of the Internet and the connection not only of personal computers, but also of a number of other devices, such as smartphones and Internet of Things devices, not only created a huge number of data sources, but also made it easier to integrate them. At the same time, better mechanisms of acquisition or creation of digital data appeared. These were also followed by an effective change in data storage capabilities. As a result, there was an unexpected change in the availability and role of data, which suddenly became widely available, surpassing

M. Kretowski, *Evolutionary Decision Trees in Large-Scale Data Mining*, Studies in Big Data 59, https://doi.org/10.1007/978-3-030-21851-5_2

the ability to effective process and analyze these data. Observing this unprecedented flood of data, the current period is increasingly referred to as the Big Data era [1].

The most common sources of Big Data are easy to point out, starting, for example, from content generated by users in social networks, which can have very diverse forms of text messages, images, or videos. In healthcare systems, next to traditional clinical or hospital data, currently, image data from various tomographic devices or large-scale genomic data (so-called omic data [2]) are generated. In transportation and logistics, vehicle positions are tracked with wireless GPS-based adapters, and parcels are identified with RFID (radio-frequency identification) badges [3]. Sensors are also placed in water pipelines, or they enable monitoring of smart grid operations and electric power consumption [4]. Scientific experiments (like the large hadron collider) or scientific simulations are a tremendous source of streaming or time series data.

As expected, there is no strict definition of Big Data, but most researchers and analysts agree on the principal data characteristics often expressed with what are called the five Vs [5]:

- Volume—This is the most obvious feature, and it refers to the vast amount of data created, generated, or sensed continually by people and machines. Every second, zettabytes of digital information are produced and shared. On Facebook alone, billions of messages are sent, and millions of new pictures are uploaded each and every day. Even storing all these data in traditional database technologies becomes impossible.
- Velocity—describes the dynamic nature of data. It encompasses both the (increasing) speed at which new data are streamed and the pace at which the data move around. Social media and high-recurrence stock exchanges are good examples of such extremely dynamic environments because they often require an on-the-fly analysis and immediate reactions.
- Variety—this relates to the inevitable large diversity of data types that must be faced. Traditionally, much of the data were neatly structured (e.g., accounting and financial data) to fit in the relational databases, but now, most of data are unstructured (e.g., text messages and conversations or images and videos from the Internet). The challenge is to harness all these data types together.
- Veracity—deals with the authenticity and credibility of the data. It is clear that for many sources, the quality and accuracy of accessible data are less controllable, so the trustworthiness of the data needs to be taken into account. In certain conditions, the volumes of data could make up for the lack of quality or accuracy.
- Value—this is often perceived as the most important characteristics of Big Data, especially from a business perspective. If the gathered data cannot be turned into something valuable, they become useless. Hence, it is essential to have a clear understanding of the costs and benefits.

Other Vs, like viability, visualisation, or vision [6], can be mentioned here, but they just emphasize the complex nature of Big Data.

It is should be noted that the amount of available data is growing very fast and the time in which the size of the collected data is likely to double soon will become several months. A lot of data that are stored will probably never be analyzed (so-called data tombs) [7]. One of the reasons behind this is that only a small fraction of available data is structured, and most of the current, more advanced analytical methods require preprocessed inputs. Classical ETL (extract-transform-load) approaches [8], which enable imposing a structure on unstructured data, are typically laborious and cannot be easily automatized. Moreover, data mining and machine learning algorithms were not originally made for exploring voluminous data. This clearly shows that an imminent need for Big Data Mining [9].

2.2 Knowledge Discovery and Data Mining

Knowledge discovery from data [10] refers to an overall process of identifying valid, novel, potentially useful, and ultimately understandable patterns or models. In fact, the idea of analyzing the gathered data and trying to take advantage of the results has been a well-known and often-used tactic for decades. Knowledge discovery has been extensively developed in the frames of statistical learning [11], pattern recognition [12], or machine learning [13]. Many interesting methods and algorithms have been proposed, and a lot of practical applications have been successfully realized, but most of these noticeable achievements have typically been limited to carefully prepared, small-sized datasets and precisely defined questions. Nowadays, with the widespread abundance of data, the perspective and expectations of data analysis methods have changed. A more holistic and autonomous process of mining nuggets in data is needed; this involves the evaluation and possible interpretation of the patterns to make the decision regarding what qualifies as knowledge. Data mining algorithms are the core of this process, but they have to account for a significantly increasing scale of problems.

2.2.1 Knowledge Discovery Process

Knowledge discovery from data is usually perceived as a multi-phase process, one that is highly interactive and iterative. The number of steps and their names can differ among authors and perspectives, but the major activities have been established and widely accepted. In Fig. 2.1 the knowledge discovery process is schematically represented.

The first activity in every knowledge discovery process should be related to understanding the application problem domain. The goals of the end-user are often expressed in business terms, and must be revealed and transformed into reasonable and achievable objectives. All available prior knowledge should be recognized, and its relevance to the problem at hand should be carefully assessed.

With this necessary insight, one can identify and analyze the available and potentially useful data sources, which can be conventional databases and data warehouses

Fig. 2.1 Knowledge
discovery process

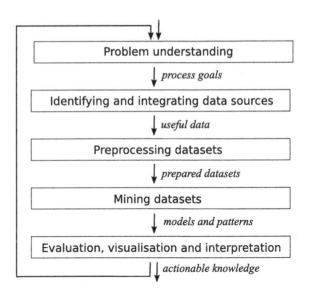

or less structured and unstructured sources in the form of simple text files, more complex visual data, or even streams. It is also possible to collect additional or supplementary data, for example, as a result of web-based queries. Another important aspect is integrating the various sources. Moreover, the correctness, completeness, and the reliability of the data must be thoroughly verified. Typically, only a limited part of the considered data sources is selected to form a target dataset that is then processed.

The next step focuses on improving the quality of the selected data and finding a direct preparation method for applying the data mining algorithms. Data cleaning encompasses the detection and possible elimination of noise or outliers. Often, real-life datasets are incomplete, so an adequate strategy must be decided for handling missing data fields. Depending on the number and cause of the data defects, the strategy can rely on a simple rejection of the deficient samples, or more sophisticated imputation techniques can be applied. Accounting for the time sequence information and known changes are necessary when the dynamic phenomena are explored. Because most data mining algorithms require a feature-based representation, the proper feature extraction and/or selection can be a key activity in data preprocessing. The dimensionality reduction (in terms of the features or instances) can also help control the time complexity of mining algorithms, especially when large-scale data come into play. In many situations, the original feature representation of the problem is not the best one, so finding a better feature space may require an application of transformation algorithms. It may also involve the discretization of continuous-valued features or recoding nominal ones.

When the datasets to be mined are ready, it is possible to proceed with actual data mining algorithms. There are many specialized algorithms, and choosing from the candidates depends primarily on the realised task, but typically, a few algorithms from the same group are used to find the right one. There are two major data mining tasks [14], namely prediction and description. In prediction, we are interested in generating a model that enables us to anticipate a value of one feature (called the target) based on the known values of other features. As for classification, the target feature is nominal, so the model assigns each new instance to one category (class) from a limited number of categories. In the case of regression, the predicted feature is numeric; hence, the predictions can be much more diverse. The model should be global in nature and could be used to predict the whole range of inputs. A decision tree is a good example of a predictive model.

Descriptive tasks are slightly different because the resulting model should try to characterize all of the data, not a single feature. A good example of such a task is a cluster analysis or segmentation, where the instances are partitioned into disjoint groups. The number of groups could be arbitrarily imposed or could possibly reflect the inherent structure of the data. A density estimation (models for the overall probability distribution of the data) and dependency modeling (models describe the relationship between features) are other descriptive modeling techniques. It should be mentioned that data mining algorithms can search for not only global models, but also local patterns or rules. Such a pattern characterizes only a certain restricted area of the feature space. Often, the discovered patterns are assessed according to their interestingness [15], which can be treated as an overall measure of the pattern's value, combining validity, novelty, usefulness, and simplicity.

As mentioned, usually, at least a few algorithms compatible with the given objectives are tested, and their parameter values are tuned to the limited subset of data. This allows for selecting the most competitive method or presenting the user with several alternative solutions.

The last step in the process is devoted to the evaluation and verification of the obtained results. The mined models and patterns can be presented to domain experts, who are hence being confronted with the current state of knowledge and understanding of the problem being addressed. To help the experts interpret the solutions, it may be important to properly visualize these solutions, especially when considering there may be alternative models or because many equally interesting patterns have been obtained. Their recommendations may refer to corrections of the current workflow or propose possible directions for further exploration. These recommendations are usually the starting point of the subsequent process iteration. Finally, if the obtained results are validated as valuable, they can be treated as new knowledge and can be used to support decisions or drive specific actions.

2.2.2 Major Components of Data Mining Algorithms

Thousands of data mining and machine learning algorithms have been introduced to date, and new ideas appear every year. When discussing the advantages and disadvantages of various approaches, it is convenient to perceive of them in a unified way. One of the most popular perspectives was proposed by Pedro Domingos [16], who defines learning as a combination of just three components: representation, evaluation, and optimization. Hand et al. [14] postulate the inclusion of a data management strategy that should be supplemented with a computation management strategy.

The first component focuses on how the searched knowledge can be represented and is closely related to the realized task. Any predictive model or pattern must be described in some formal language, and it determines the underlying structure or functional forms that are being sought from the data. Only solutions that can be expressed within the adopted formalism can be learned. This searched space is called the hypothesis space of the learner. For the most typical predictive task, the searched model may take the form of, for example, an artificial neural network, graphical model, a set of rules, or a decision tree. It can also be linear or non-linear, instance based, or hyperplane based.

The second element requires defining an evaluation function, which is necessary to distinguish good solutions from the weak ones and to choose a good set of values for the parameters of the model. The evaluation function can be called the objective or scoring function. It precisely reflects the utility of a particular model, which could be seen as the true expected benefit, but in reality, it is rather intractable. Instead, the general scoring functions are applied, and these typically confront the considered model or pattern with the mined dataset. Typical examples in prediction tasks are the misclassification rate for classification problems and the sum of squared errors for a regression.

When the solution representation and evaluation function have been decided upon, the last missing component is the search or optimization mechanism. The search mechanism goal is to determine the structure and some parameter values that can achieve the best value of the score function. The task of finding interesting structural patterns (such as decision rules or trees) is usually seen as a combinatorial search problem that can be solved using heuristic search techniques (like greedy and beam searches or branch-and-bound searches). However, the task of finding the best parameter values in the models (e.g., fitting coefficients in a linear regression model) is usually considered to be a problem of continuous optimisation, which can be solved with unconstrained (like a gradient descent or conjugate gradient) or constrained (like linear programming or quadratic programming) techniques.

The last component encompasses the ways in which the data are stored, indexed, and accessed. It is especially important in the context of large-scale data mining where the assumption that the whole dataset is loaded into an operational memory of a single machine could be unrealistic. The necessary supporting mechanisms could employ indirect access or, more probably, distributed and/or parallel processing.

Here, the aforementioned vision of data mining algorithms as optimization and search processes is clearly coherent with the evolutionary approach discussed in Chap. 1, and the development of evolutionary data mining methods seems to be not only an expected but desirable idea.

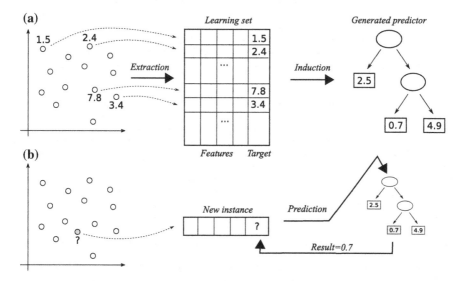

Fig. 2.2 Two phases of predictive modeling: **a** learning, **b** predicting

2.2.3 Classification and Regression

As shown, prediction is one of the most common data mining tasks, and depending on the type of the target feature, a classification and regression are applied. In this book, I will focus on the prediction task only, and both types of predictive modeling will be studied. The models, which are used to predict an unknown value of the target feature, are generally called predictors, but when it comes to a classification, a more precise term, specifically called as classifier, is usually prefered.[1] In all predictive approaches two phases are normally distinguished. In the first phase, a predictor is learned (generated or induced) based on the training dataset. In the second phase, the generated predictor can be used to anticipate (with some certainty) the value of a target feature of new instances on the condition that the values of the needed features are given as inputs for. From the available features in the training dataset, the subset of relevant features required by a predictor is typically selected during the learning. For example, decision trees usually request only a small fraction of original features, whereas other methods, like artificial neural networks, exploit all the features. In Fig. 2.2, two phases of predictive modeling are depicted.

The learning dataset $L = \{\mathbf{x}_1, \mathbf{x}_2, ..., \mathbf{x}_M\}$ is composed of M instances (objects or examples) belonging to the problem domain U ($L \subset U$). Every instance is described by N features (attributes) A_j. Any feature A_j is a function defined on U in the form of $U \rightarrow V(A_j)$, where $V(A_j)$ is a set of feature values. A value of j feature (A_j) of i instance ($\mathbf{x}_i$) can be denoted as $x_{ij} = A_j(\mathbf{x}_i)$. There are two types of features that should be distinguished:

---

[1] In a regression, the independent features are called regressors.

- Nominal (discrete-valued) features—its domain is a finite set $V(A_j) = \{v_j^1, ..., v_j^{S_j}\}$, where $S_j = |V(A_j)|$. A typical example of a nominal feature is *HairColor*, and the set of possible values can be defined, for example, as $V(HairColor) = \{blond, brunet, ginger, gray\}$. Nominal features can be well-fitted for certain algorithms, but they can also pose some problems for other methods, and some coding may be necessary. The simplest nominal feature has only two values (e.g., *true* and *false*), and such a binary feature is often represented as 1 and 0 (or 1 and -1). For certain nominal features (called ordinal), it is possible to define an ordering (ranking) of possible values. This allows for coding them with natural numbers and treating them in a similar way as numeric features.
- Numeric (continuous-valued) features—typically $V(A_j) = \Re$, or the possible values are limited to an interval $V(A_j) = [l_j, u_j]$, where l_j and u_j are the lower and upper limits, respectively. A good example of a continuous-valued feature is *Height* with $V(Height) = [0, 250]$. If only nominal features are accepted by an algorithm, a discretization [17] of numeric features needs to be performed.

Instances are labeled by a target feature TF in prediction problems. Every instance $(\mathbf{x}_i)$ is associated with one value of the target $(tf_i = TF(\mathbf{x}_i))$. In classification problems, the target feature is nominal and is usually called the class or decision (can be denoted as $d(\mathbf{x}_i)$). In regression problems, the dependent variable is numerical. Formally, a predictor κ is a function from a feature space into a set of target values:

$$\kappa : V(A_1) \times \cdots \times V(A_N) \to V(TF). \tag{2.1}$$

A value of the target feature proposed by a predictor κ for a new instance $\mathbf{x}$ can be denoted as $\kappa(\mathbf{x})$. In the case of an ideal predictor, $\kappa(\mathbf{x}) = TF(\mathbf{x})$ for all instances from the problem domain. In practice, such a situation is rather unlikely.

2.3 Decision Trees

Decision trees [18] are one of the most popular forms of knowledge representation in data mining and knowledge discovery [19]. The obtained results are usually competitive, and both the processes of learning and predicting are understandable, especially when compared with complex, black-box approaches like ensembles [20] or deep-learned convolutional neural networks [21].

Decision trees are hierarchical, sequential structures (acyclic, directed graph) composed of nodes (vertices) and branches (directed edges) connecting nodes. The first node is called the root, and it is the only node without a predecessor (all other nodes in the tree have exactly one predecessor). Two types of nodes need to be distinguished: internal nodes (non-terminal nodes) and leaves (terminal nodes). Any internal node has at least two or more descendants, whereas a leaf has no descendants. The simplest decision tree consists of just one node (a leaf). With every internal node, a corresponding test (with two or more outcomes) is associated. The number of test

Fig. 2.3 Decision tree
fundamentals

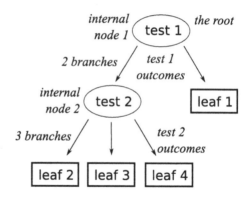

outcomes should be equal to the number of branches coming from that node because
each test outcome is linked to one branch. Terminal nodes do not have any tests, but
they are responsible for prediction: in a classification tree, the leaves are associated
with classes, whereas for a regression problem, a leaf predicts a numerical target.
The most basic components of a decision tree are presented in Fig. 2.3.

Even if the creation of a decision tree may seem mysterious for novices, its usage
is really simple and intuitive. For any new instance (object), we have to traverse it
through the tree starting from the root node until reaching one of the leaves. This
means that at first, a test associated with the root node is performed.[2] According
to this test result, one branch linked with an adequate test outcome is chosen, and
the object is moved down to the corresponding node. The procedure is recursively
repeated. If the node is not terminal, the next test is checked, and the object is further
moved down. If finally, a reached node is a leaf, a proper prediction for the instance
can be assigned based on the target value (or model) associated with the leaf.

This resembles the human way of reasoning, which is typically gradual and pro-
gressive. The most important arguments and conditions are analyzed first, and then,
based on the results, more specific but only adequate elements are considered. This
way of proceeding is fast and consistent. Although it requires a strict ordering of
considered factors; without a doubt, it is easier than, for example, defining precise
weights for multiple criteria. Every prediction is well justified because the tests on the
traversed path from the root to the leaf can be always visualized or presented as the
"if-then" rule. More generally, any decision tree can be converted into a set of rules
(e.g., [22]). However, the tree guarantees a prediction for any instance, which is not
the case for a set of rules. In addition, a decision tree inherently performs the feature
selection because only a limited number of features is used in tests, and only the
values of these features are necessary for proposing a prediction of a new instance.
A more detailed description of the advantages and disadvantages of decision trees
can be found in [23].

---

[2]We assume that the tree has at least one internal node and is not reduced to just one leaf.

2.3.1 Tree Types

Different types of decision trees can be distinguished by taking into account the test types used in internal nodes:

- Univariate test—just one feature value is verified in the test. The most common tests are inequality tests, which can be denoted in the form $(A_j < th)$, where th is the given threshold value. There are two outcomes of this type of a test: an instance is directed to the left or right branch depending on the value of A_j feature: whether it is lower than th or not. Concerning the simplest test that is based on a nominal feature, it assigns a separate outcome to any possible feature value. If the number of feature values is high, this may pose some problems. Feature values can also be grouped, and such groups can be associated with branches. This is called an internal disjunction. If only two non-empty groups are considered, this type of binary test can be seen as the equivalent of an inequality test.
- Multivariate test—this requires more than one feature in the test. The most popular example is an oblique split, which is based on a given hyperplane $H(\mathbf{w}, \theta)$ representing a linear combination of all feature values:

$$H(\mathbf{w}, \theta) = \{\mathbf{x} : \langle \mathbf{x}, \mathbf{w} \rangle = \theta\}, \tag{2.2}$$

where $\langle , \rangle$ is an inner product. A hyperplane divides a feature space into two parts, and the corresponding test checks if the instance $\mathbf{x}_i$ is located on the positive or negative side of the hyperplane:

$$\langle \mathbf{x}_i, \mathbf{w} \rangle - \theta > 0. \tag{2.3}$$

Another special case is a multi-test, where a test is composed of a few univariate, binary tests. The results of a multi-test can be determined by a majority voting system.

Typically, one type of test is used in the whole tree, and such homogenous tree structures are often named according to the test type, for example, univariate decision trees with only univariate tests or oblique trees with only oblique splits. Oblique trees can also be called perceptron trees [24]. In Fig. 2.4, examples of the most common tree types are presented. The term mixed decision tree was proposed by Llora and Wilson [25], and it corresponds to heterogenous trees with various types of tests. They are also called omnivariate decision trees [26].

Univariate trees are obviously the most popular type of tree. Among the hundreds of solutions proposed over the past fifty years [27], it is necessary to mention at least the family of systems developed by J. Quinlan (ID3[3] [28] and C4.5 [22]) and CHAID (CHi-squared Automatic Interaction Detection) [29] proposed by G. Kass. They are still reference works for many other approaches.

---

[3]ID3 stands for Iterative Dichotomiser 3.

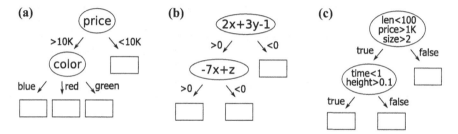

Fig. 2.4 Univariate, oblique, and multi-test decision trees

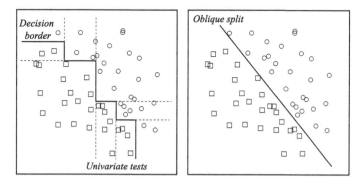

Fig. 2.5 Staircase effect

In terms of interpretability, a small univariate tree is hard to beat, but when the size of a tree grows and reaches more than several dozen nodes, understanding the tree becomes difficult. For certain problems, relying on more complicated oblique tests can bring less-complicated trees. From a geometrical point of view, univariate tests are axis-parallel, whereas splitting hyperplanes into oblique tests can be arranged arbitrarily. In the case of a problem where decision borders are inherently not axis-parallel, applying only univariate tests could lead to their approximation through the use of very complicated stair-like structures [30]. This situation is known as the staircase effect. In Fig. 2.5, an example of a classification dataset, where many univariate tests can be easily replaced by a single oblique test, is shown.

On the other hand, even a single oblique test with a lot of features may not be easy to interpret. This is why feature selection methods are typically applied to eliminate as many coefficients from the splitting hyperplanes. A prominent example of this approach is CART (Classification and Regression Tree) [31], which permits both univariate and oblique tests. It can search for a linear combination of the continuous-valued features and also simplify it using a feature elimination procedure. However, the CART system clearly prefers univariate tests and as a result very rarely utilizes more complicated tests. The approach has been extended in many ways. OC1 (Oblique Classifier 1) [32] combines deterministic (hill-climbing) and randomized procedures to search for a good tree. In [33], the *Linear tree* system is proposed, and

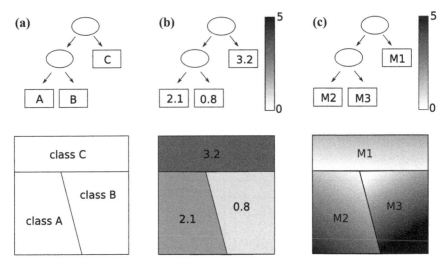

Fig. 2.6 Classification, regression, and model trees

it combines an univariate tree with a linear discrimination by means of constructive induction. At each node, a new instance space is defined by inserting new features that are projections over the hyperplanes, which are given by a linear discrimination function.

Decision trees can be dedicated to classification problems when the nominal predictions are considered and to regression problems when the predicted feature is a continuous-valued one. As a result, *classification* and *regression trees* are distinguished. In classification trees, every leaf is associated with a decision (single value of a nominal predictive feature), and usually, this value is chosen with a majority voting system. Voting is based on the learning instances that have reached the leaf. In the case of a regression tree, the situation can be a little bit more complicated. In the simplest situation, a leaf is associated with an average value of the target feature (calculated using the learning instances in the leaf). In a more advanced situation, in every leaf, a model is created and stored. When a new instance reaches one of the leaves, the corresponding model is applied, and the prediction is calculated. This type of regression tree is called a *model tree*. There are different types of (typically linear) models that can be applied, ranging from a simple linear regression (where only one feature is taken into account) to a multiple linear regression. In Fig. 2.6, three types of decision trees are presented.

As for regression problems, CART minimizes the weighted sum of the variances of the target in the created subsets of the training data. It uses the mean values of the target in the leaves. M5 [34], which is similar to CART, minimizes the weighted sum of the standard deviations, but it fits a linear regression model for each leaf of the model tree. Moreover, in the M5 system, a smoothing process to improve the prediction accuracy is introduced. The smoothing tries to reduce sharp discontinuities that could occur

Fig. 2.7 An example of a
smoothed prediction by a
model tree

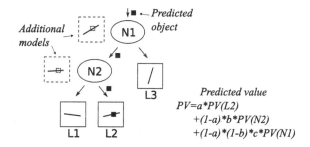

Predicted value
$PV=a*PV(L2)$
$+(1-a)*b*PV(N2)$
$+(1-a)*(1-b)*c*PV(N1)$

between adjacent models in the leaves. It requires fitting additional models in all non-terminal nodes, and these models are applied to adjust the predictions calculated using only the models in the leaves. However, instead of using all features, every additional model is restricted to the features that are referenced by tests or linear models somewhere in the subtree of this node.

When a smoothed model tree predicts a target value for an instance, the beginning is classical: the instance crosses a path from a root node to a leaf, and an associated model is deployed. Next, the path to the root node is traced back, and additional models in internal nodes are stepwise used to smooth the original prediction while decreasing influence on the target value. In a formula for a recursive calculation (adjusting at each step), the predicted value PV is as follows:

$$PV(S) = \frac{n_x \times PV(S_x) + k \times M(S)}{n_x + k},\tag{2.4}$$

where S_x is a current node where the prediction is known, S is its parent (internal node on a path to a root), n_x is the number of training instances at S_x, $M(S)$ is the predicted value calculated from an additional model at S, and k is a smoothing constant (default $k = 15$). In Fig. 2.7, an example of predicting with a smoothed model tree is presented.

The smoothing mechanism has the greatest effect on a target's value prediction when some models in the lower parts of a tree are fitted with few training instances, and/or the models along the path predict very differently.

The HTL algorithm [35] also applies the CART-like splitting criterion, but is not restricted to linear models within the leaf nodes.

In the context of Big Data Mining, we are interested in the learning (automatic generation) of the decision trees for a given classification or regression problem by using the available data. The number of possible trees, even for an uncomplicated problem is huge. Various preference criteria can be defined, but typically, a user is interested in a decision tree generalizes well and can predict robustly. It is rather intuitive that the decision tree's induction algorithms are highly computationally demanding. And this expectation was theoretically confirmed in a few papers. For example, Hayfil and Rivest [36] show that constructing a minimal binary tree with respect to the expected number of tests required for classifying an unseen instance is NP-complete.

2.3.2　Top-Down Induction

One of the most fundamental algorithm design principles is "divide and conquer" [37], which postulates solving a difficult problem by recursively breaking it down into sub-problems of the same or related type until these become simple enough to be solved directly. The partial solutions are then combined to give a final solution of the original problem. A top-down induction of decision trees is a direct realization of the divide and conquer principle. Currently, this is the most popular heuristic applied to automatically generate decision trees from a training dataset [38]. The clones of the seminal top-down algorithms like CART [31] or C4.5 [22] are available in virtually all commercial or academic data mining systems. These algorithms are relatively easy to understand, and the learning is fast. The quality of the results is usually not bad, and the generated models are typically much easier to interpret than black-box solutions (e.g., artificial neural networks).

The top-down algorithm works as follows: First, a root node is created, and a full training dataset is associated with this node. Then, a recursive procedure can be launched, which tries to hierarchically divide the training data. At every step of the algorithm, two alternative situations are possible: a node can be marked as a leaf, and then the induction is finished, or an effective test is found, and then, the tree growing process continues. The stopping rule is usually linked with available data homogeneity according to the target. For example, in a classification, if all the training instances in a node have the same decision, further splitting does not make sense. It is also impossible to find a split if there is only one available training object. Furthermore, if the number of available objects in a node is small, it is also reasonable to break these down because the generalization property of a test found on such a limited basis will be poor. Typically, the algorithms impose a requirement of at least five training instances in an internal node.

Finding a good test for a node is clearly a crucial property of the algorithm. For the simplest test types, all possible tests are enumerated, and then, the goodness of corresponding splits is evaluated according to a certain optimality criterion (a few classical criteria will be discussed later in this section). For example, if the most common inequality test is searched, the procedure iterates through the attributes and the possible thresholds derived from the attribute values in the training data. For a more complex test, like an oblique one, this strategy is not possible because enumerating all possible candidates is impossible. In such a situation, a more advanced search strategy (based, e.g., on a meta-heuristic) is usually applied. When the required test is found, new nodes could be created. The number of newly created nodes is equal to the number of the test's possible outcomes, and one branch connects a parent node with every new node. Each branch is permanently associated with one test outcome. Now, the training instances can be pushed down into the newly created nodes. For each training instance in the current node, the test is performed, and according to the test's results, this instance may be located in one of the new nodes. When all the training instances from the current node are relocated, the next level of a recursion can be performed in the newly created nodes. It should be noted that in each step of the recursive procedure, the number of available data in a node decreases, and the nodes are more likely to become leaves. A prediction associated with a leaf or calculated in a leaf always depends on a subset of the training data, that reached this terminal-node. In Fig. 2.8, an illustrative example of a top-down classification tree induction for a two-dimensional problem with four classes is presented. It can be observed how the decision boundaries between the learning instances representing different classes are formed hierarchically.

Post-pruning

Decision trees generated with the aforementioned algorithm can successfully retrieve the predictions of virtually all learning instances, but in fact, it is not an actual aim of the learning. We are primarily interested in getting a predictor that will be able to properly forecast if there are previously unseen instances. Unfortunately, it is much more difficult because the problem of overfitting to the learning data appears [12]. This problem is observed for most of the algorithms that learn a predictive model based on the dataset. Overfitting manifests in a very good prediction for the learning instances but in much worse predictions for new instances. The phenomenon can be explained by not enough ability to generalize. The algorithms are typically too concentrated on perceiving the details of the available data, and they work more as memories, not models.

As for a decision tree, it means that too many splits are used and much too detailed leaves are being created (with only a small number of training instances). It would seem that the problem can be solved by proposing more conservative stopping criteria, but in fact, it is difficult to introduce a universal mechanism that will work well in diverse situations. It has become widely accepted [38] to first induce a possible overgrown tree and then apply a post-pruning to reduce the tree size and increase its generalization. This approach is slightly more computationally complex, but it offers a more reliable analysis of the situation and eliminates the risk of stopping too early.

Most of the post-pruning algorithms work in an analogous way. They analyze the tree generated by the top-down induction in a bottom-up fashion and check if any replacement of a node by a leaf improves the prediction accuracy. The initial checks are applied to small parts of the tree, but subsequent attempts include larger and larger parts. In some of the algorithms, it is also possible to move up one of the branches and to replace a subtree by a part of this subtree. The most crucial element in every algorithm is a way to get at the prediction accuracy estimation. Obviously, this cannot rely only on the training instances because any pruning will not improve their predictions. In Fig. 2.9, an example of a generic post-pruning algorithm is presented.

Among many post-pruning techniques [39], there are three classical methods:

- Reduced-error pruning—it requires a detached part (usually 25% or 33%) of a learning dataset that is used only for an error estimation during pruning, not for

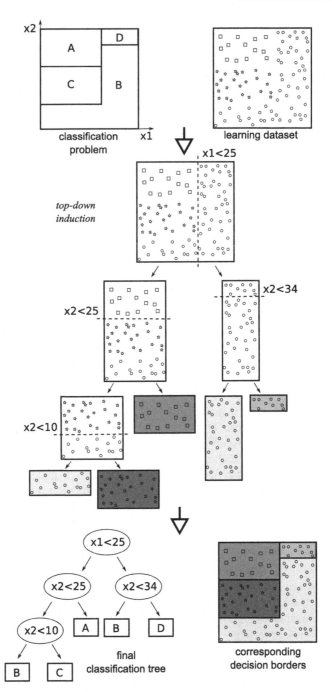

Fig. 2.8 An example of a top-down induction of univariate tree (a classification problem)

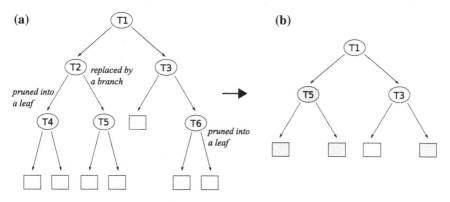

Fig. 2.9 An example of post-pruning: **a** an input tree, **b** a pruned tree

the induction. From a theoretical point of view, it is a well-justified method, but in practice, this method is applied reluctantly because not all available data can be directly utilized for the induction. It is especially hindering when the learning dataset is relatively small.

- Pessimistic pruning—the main idea is the systematic increase of the overoptimistic error estimates derived from the entire training dataset. The method was originally introduced by J. Quinlan [40] in the ID3 system, and it simply adds 0.5 to the number of errors in each leaf. Additionally, a special constraint is imposed (one standard error of difference) when deciding on every subtree pruning. In C4.5 [22], another heuristic rule for the pessimistic error calculation is proposed. It treats the learning subset in a node as a sample from a binomial distribution, and for a given confidence level, it adopts an upper end of the confidence interval as an error estimate.
- Cost-complexity pruning—the method is proposed in the CART system [31], and it works in two phases. In the first phase, starting from the largest tree, a sequence of trees of decreasing size is constructed. The last tree in the sequence is composed of a single leaf. The trees are chosen to minimize the cost-complexity criteria, which takes the form of:

$$CC(T) = Error(T) + \alpha * Size(T), \qquad (2.5)$$

where $Error(T)$ is the tree T error estimated on the learning dataset, $Size(T)$ corresponds to the tree T size, and α is a non-negative complexity importance factor. By adjusting the α term, a variation of the tree sizes can be obtained. In the second phase of the algorithm, all trees in the sequence are assessed with a cross-validation procedure, and the best tree is returned. This pruning method is especially interesting because the fitness function in the evolutionary induction will also be based on the trade-off between accuracy and complexity.

Goodness of Split Optimality Criteria

As noted, the locally best test can be chosen from a set of enumerated proposals based on the adopted goodness of the split criteria. Typically, an impurity measure is used to characterize the learning data before and after applying the test. The test that maximizes the impurity reduction is preferred. For a given test s with ns outcomes and the learning data L, the impurity difference $\Delta(s, L)$ is defined as follows:

$$\Delta(s, L) = I(L) - \frac{1}{|L|} \cdot \sum_{i=1}^{ns} (|L_i| \cdot I(L_i)), \tag{2.6}$$

where $I(L)$ stands for the impurity measure of L, and L_i corresponds to the subsets generated by the split s.

In classification problems, the impurity should be minimal when all the instances belong to the same class and maximal when the classes are represented equally among the analysed data. In the ID3 system [28], entropy as a measure of class diversity is applied, and the resulting criteria are known as the information gain. The entropy of the learning data L can be formulated as:

$$Entropy(L) = - \sum_{c_i \in C} \frac{|c_i(L)|}{|L|} \cdot log_2 \frac{|c_i(L)|}{|L|}, \tag{2.7}$$

where $C = \{c_1, ..., c_K\}$ is a set classes (targets), and $c_i(L)$ is a set of instances in L with a target equal to c_i.

It is well-known that the information gain has a bias toward tests with more outcomes. This does not matter for inequality tests with two outcomes, but for tests devoted to nominal features, it can be a serious problem. Hence, a criteria adjustment is proposed in C4.5 [22]. The information gain is divided by a split information that gives the gain ratio, which is applied for the test assessment. The split information $SI(s, L)$ is a kind of a penalty for more outcomes and is defined as follows:

$$SI(s, L) = - \frac{1}{|L|} \cdot \sum_{i=1}^{ns} |L_i| \cdot log \frac{|L_i|}{|L|}. \tag{2.8}$$

In the CART system [31], the Gini index of diversity is applied, and it can be defined as follows:

$$G(L) = - \sum_{i \neq j} \frac{|c_i(L)|}{|L|} \cdot \frac{|c_j(L)|}{|L|} = 1 - \sum_{c_i \in C} \left(\frac{|c_i(L)|}{|L|} \right)^2. \tag{2.9}$$

The impurity reduction is not the only possible approach that can be applied to select a test in a node. For example, in CART, a twoing rule is proposed as an alternative, but it can be used only for binary tests. Another numerous group of systems takes advantage of statistical independence tests. CHAID [29] is a well-known representative, and it applies the χ^2 statistic to verify if the split of the learning instances imposed by a given test is independent from a class attribute.

In regression problems, the node impurity can be measured by the sum of squared residuals (the differences between the predictions and the actual values) over all instances in the node. These criteria are usually called *Least squares*, and the residual sum of squares (RSS) for the training set L can be formulated as follows:

$$RSS(L) = \sum_{i=1}^{M}(tf_i - T(\mathbf{x}_i))^2, \tag{2.10}$$

where $T()$ is the prediction of the tree T.

As an alternative, *Least Absolute Deviation* can be used, enabling a reduction of the sum of the absolute value of the residuals:

$$LAD(L) = \sum_{i=1}^{M} |tf_i - T(\mathbf{x}_i)|. \tag{2.11}$$

It can be more robust in certain situations than the first measure because it is less sensitive to outliers.

It is widely accepted that in typical, simple problems, the top-down induction of decision trees ensures generally good predictive accuracy. Moreover, a greedy induction is relatively fast, especially when the number of considered candidate tests is moderate. It is also well-known that top-down-induced trees, even after post-pruning, are often overgrown, and when a tree is composed of dozens or hundreds of nodes, its interpretability is questionable.

2.3.3 Going More Global

Deficiencies of the greedy top-down induction come directly from their step-wise nature, where only locally optimal decisions can be performed. And in many situations, such a scheme does not lead to globally optimal solutions that would be understood as both accurate and simple decision trees. The problem can be easily revealed with the help of two examples (Fig. 2.10). In both cases, let us analyze two-dimensional, artificially generated datasets with analytically defined decision borders. The first one is a classification problem called *chess2x2* [41], which could be seen as a kind of generalisation of the well-known XOR problem. An eligible classification tree should be composed of three internal nodes and four leaves. However, one of the most renowned data mining system—C4.5 [22]—is not able to find any good solution and returns a single leaf with a default class. As with most of classical systems, it analyses the features independently, one by one. For every possible candidate split and typical optimality criterion (based on the gain ratio or Gini index), an even distribution of the classes is preserved, and a gain is negligible, which stops

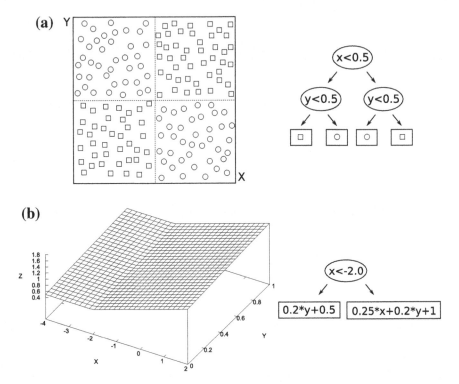

Fig. 2.10 Examples of classification (**a**) and regression (**b**) problems where typical top-down inducers fail. On the right are the corresponding optimal decision trees

the top-down induction. It is clear that a more global system could better find an appropriate tree.

The second example pertains to a regression problem with two distinct regions where a target feature is linearly dependent on one or two independent features (*SplitPlane3* dataset from [42]). The optimal model tree has only one internal node with a partition at threshold $x_1 = -2.0$ and two linear models in leaves, as presented in Fig. 2.10 on the right. This simple problem is nevertheless insidious, even for the best-known top-down greedy inducers. The M5 algorithm [34] is trapped in a local optimum because it first splits at $x_1 = -1.2$. Such a not-optimal partition in the root node increases the tree complexity, and the final tree has four internal nodes and five leaves with three different linear models. For this dataset, the CART system [31] generates an even more complicated tree with a nonoptimal split ($x_1 = -0.44$) at the root node.

The problem of the inadequacy of greedy induction, especially for difficult concepts, was pointed out early, and several attempts were performed to find a solution [23]. One interesting proposition that tries to search more globally is called APDT (*Alopex Perceptron Decision Tree*) [43] and is proposed by Shah and Sastry. It generates binary oblique classification trees with a top-down method, but evaluates hyper-planes by taking into account possible interactions with the splits in the next level of the tree. Such a one-look-ahead procedure is clearly more computationally complex, but it significantly improves the generalization ability of the algorithm. A similar idea was implemented in the LLRT (*Look-Ahead Linear Regression Tree*) [44]. The system performs a near-exhaustive evaluation of the set of possible splits in a node. The splits are rated with respect to the quality of the linear regression models and are fitted to the training objects in the resulting branches.

Even recently, Wang *et al.* [45] have proposed the TEIM (*Tsalis Entropy Information Metric*) approach, which is based on two-term Tsalis conditional entropy but with a tunable parameter and a new tree construction method. The authors claim that the two-stage tree construction method is influenced by previous attributes and class labels, which reduces the greediness and avoids local optimum, to a certain extent.

It should also be mentioned that metaheuristics other than evolutionary computation can be applied in the data mining based on decision trees. For example, ant colony optimization-based methods are proposed in [46] for both decision tree and ensemble learning.

2.4 Evolutionary Approaches for Decision Tree Induction

Evolutionary computations, because of their flexibility and reliability, early on attracted the attention of data mining researchers [47]. They were successfully applied in various knowledge discovery tasks and methods like, for example, feature selection, clustering, or decision rule learning. In this section, only evolutionary approaches dealing with a decision tree induction will be discussed.

There are three main ways of applying evolutionary computations to a decision tree induction. In the first, an evolutionary search is applied in internal nodes for finding optimal (usually oblique) splits during the top-down induction. The second group is also linked to the top-down induction, but this time, an evolution is used to choose among different induction methods and to search for the optimal parameters of them. The last group, which is the most closely linked to the approach described in this book, encompasses methods that use a global induction, where a full tree is searched during an evolution. Below, the most prominent examples of evolutionary decision tree inducers are discussed.

2.4.1 Evolutionary Split Searching

The BTGA (*Binary Tree-Genetic Algorithm*) system [48] was one of the first solutions. It uses a standard genetic algorithm with a binary representation to search for the optimal splitting hyperplane in every internal node. A similar approach is presented in [49], where the original OC1 system [32] was extended by applying two classical algorithms: evolutionary strategies (1+1) and a standard genetic algorithm. The main drawback of these methods is their computational complexity, which resulted in long induction times. As a result, Ng i Leung [50] tries to speed up an evolutionary search for linear splits by using additional k-D tree structures, created before every search in the non-terminal node has been launched. Another example of the oblique tree generation method is presented by Tan and Dove in [51]. Their system uses a fitness function that is based on the Minimum Message Length (MML) principle in a simple evolutionary strategy. Finally, the author of this book investigated this type of the evolutionary induction. In [52], an evolutionary algorithm based on a dipolar criterion function with specialized differentiation operators was proposed for finding splitting hyperplanes.

2.4.2 Evolutionary Meta-Learning

A classical top-down univariate tree inducers are in general fast, especially for middle-scale problems; with this method, some researchers have tried to exploit the meta-learning approach [53]. Barros et al. propose *Hyper-heuristic Evolutionary Algorithm for Designing Decision Tree* (HEAD-DT) [54]. The authors apply an evolutionary design to decision tree top-down induction by assembling the algorithm from various heuristic design components (like, e.g., splitting and stopping criteria or a pruning method). The system is learned on the meta-dataset composed of many standard datasets, and the classification results averaged on the group are used as a fitness function. The approach has been successfully applied to microarray gene expression data. Recently, Karabadji et al. [55] present another meta-heuristic approach based on a simple genetic algorithm that searches for a given dataset, the best combination of a subset of attributes, proportion of training and testing examples, top-down induction algorithm (from four decision tree algorithms available in WEKA [56]), and some secondary parameters. The fitness function is a weighted sum of the precision, trust, and simplicity of the found combination. As expected, such a method outperforms the state-of-the-art decision tree algorithms, like C4.5 [22] and CART [31].

2.4.3 Global Evolution

In the third group, the global approaches, the tree structure and test are searched simultaneously in a single run of the algorithm. This type of research originates in the genetic programming [57] community where the tree representation of evolving programs is typically used and a relatively simple adaptation to decision trees is possible. In [58], John Koza, the pioneer of genetic programming, sketches an adaptation where LISP S-expressions, corresponding to decision trees with only nominal tests, are evolved. The idea has been further developed by others in [59], where a specialized fitness function based on the stochastic complexity and standard genetic operators enable the evolution of programs representing univariate trees. Tanigawa and Zhao investigate the induction of binary classification trees with inequality tests, and they propose two methods for the size reduction of a resulting tree [60]. On the other hand, Bot and Langdon [61] apply genetic programming to generate classification trees with limited oblique tests. In their system, up to three features could be used to define a single split.

An interesting cycle of papers was published by Folino and co-workers. In the first [62], they apply a cellular genetic programming-based approach to simulate the evolution of simplified trees but by using only nominal tests. The method uses cellular automata as a framework to enable a fine-grained parallel implementation through the diffusion model. In the next paper [63], the approach is extended by the simulated annealing schedule to decide the acceptance of new solutions. In another paper [64], the authors propose a new fitness function based on the J-measure.

Another approach based on genetic programming is presented by Kuo et al. [65]. They decide to simplify a classification tree into a ranked tree that corresponds to an ordered list of decision rules. As a result, specialized genetic operators are proposed, for example, to eliminate repeated or more specific rules. A fitness function is proportional to the reclassification accuracy and inversely proportional to a given power of a normalized classifier size.

In [66], the authors study a GP-inspired classification tree induction with multi-interval tests instead of typical inequality tests for continuous-valued attributes. The initial thresholds are calculated based on the MML principle. The number of thresholds in each test evolves because of specialized mutation and crossover operators. Here, any N-interval test is equivalent to the cascade (subtree) of (N-1) the inequality tests.

However, a direct porting of the decision tree construction into a genetic programming framework is not an easy task because the common tree representation is only seemingly facilitated. Most of the approaches are limited to defining the appropriate fitness function and to selecting the acceptable functors and terminal symbols, which it turns out is not enough for a competitive and robust induction.

One of the first global systems that was not genetic programming-based, is proposed by Papagelis and Kales [67]. Their GATree system uses only tests with nominal features, and its weighted fitness function takes into account both reclassification accuracy and the tree size. Interestingly, the mutation is not uniformly applied to tree nodes. In [68], they show that the partial fitness calculations could be reused efficiently if the training instance indices were stored in each node. Only modified by genetic operators, the tree parts became stale and needed recalculations. This results in a substantial induction time reduction, especially when higher memory complexity is not a constraint.

In GAIT (*Genetic Algorithm for Intelligent Trees*) [69, 70], which is another genetic algorithm-based system, an initial population of binary trees is created by applying C4.5 [22] to randomly drawn small subsamples of the learning data. There are two simple genetic operators, and individuals are evaluated according to the classification accuracy estimated on the separate validation subset. Here, only tests from the initial trees can be used in the obtained tree. The approach has been successfully tested on relatively large datasets (e.g., 440 000 instances).

Sorensen and Janssens [71] investigate automatic interaction detection (AID) techniques for evolving fixed-size classification trees with only binary predictors. They develop a genetic algorithm that tries to find the optimal content of a tree. Interestingly, the algorithm maintains a list of solutions with the highest fitness, and this list is finally presented for an analyst to choose the one tree that best meets the analyst's requirements. This approach can be seen as a step toward a multi-objective evolutionary method, where the final solution is selected from a Pareto front.

It is also worth mentioning another global induction system called GALE (*Genetic and Artificial Life Environment*) [72], which is based on a fine-grained (cellular) parallel evolutionary algorithm. Besides typical univariate and oblique tests, the system permits specific multivariate tests based on a prototype (an artificially defined instance) and an activation threshold. Here, only the squared reclassification accuracy is used as a fitness function. And only simple differentiation operators (one-point crossover from genetic programming and random test modifications) are applied.

In [73], the authors investigate an application of genetic algorithms for finding compact classification trees in a case where the dataset is composed of only binary features. Interestingly, they propose the sum of the depth of the leaves for each learning object (corresponding to the number of tests from the root to a leaf) as a measure of the tree complexity (and a fitness function). Binary trees are encoded based on the breadth-first traversal as a fixed-size list of attribute indexes.

Most attention has been focused on classification trees, but there are also interesting works focused on solving regression problems. One of the first applications of an evolutionary approach to the induction of regression trees is presented in [74]. The TARGET system evolves univariate trees with simple genetic operators: swapping subtrees or tests as a crossover; randomly changing a test; or swapping the parts of the same tree as a mutation. Additionally, a transplant operation is introduced, adding a new randomly generated tree to the population. The Bayesian information criterion (BIC) [75] is applied as a fitness function. In another paper [76], the authors adapt their system to construct classification trees and incorporated an option of splitting rules based on the linear combinations of two or three variables.

Another simplistic system for a univariate regression tree induction called STGP (*Strongly Typed Genetic Programming*) is described in [77]. Only the prediction error (RMSE) on the training data is used a fitness function, whereas the tree size is limited by the maximum depth parameter. It applies just basic variants of mutation and crossover from genetic programming.

The global evolutionary approach can also be applied to more complex, model tree induction. In [78], the authors present the E-Motion system, which searches for a binary, univariate tree with linear models in the leaves. An initial population is composed of the stumps (trees with only one internal node and two leaves), minimizing a standard deviation of a single attribute. Linear models are generated according to the M5 algorithm. The system offers two multi-objective fitness functions (based on the weight formula or lexicographic analysis), both considering the prediction errors (RMSE and MAE) and tree size (number of nodes) with arbitrarily chosen weights or tolerance thresholds. Basis variants of the genetic operators are used: standard one-point crossover and pruning of an internal node or expanding a leaf as mutations. The most advanced model tree representation with nonlinear models in the leaves is implemented in GPMCC, which stands for the GP approach for mining continuous-valued classes [79]. The system applies two evolutionary techniques at distinct levels: genetic programming for evolving the tree structure and genetic algorithms for the evolution of polynomial expressions (models) [80]. The fitness function is based on the extended form of the adjusted coefficient of determination. Among classical forms of a crossover, the system uses non-standard mutation strategies, where, for example, the node with the higher MSE is modified with the higher probability.

In [81], Sprogar investigates the role of the crossover operator in the genetic programming-based induction of decision trees. He introduces a novel, context-aware variant of the crossover, where the context is defined by the training samples at the crossover point. In the proposed (so called prudent) operator, the subtree can only be moved from one tree to the other when at least one training sample is common in both contexts. Moreover, the crossover point can be selected using a random choice or (with a given probability) by using the highest information gain. The experimental validation shows that such an informed variant of the crossover could be profitable in terms of its generalization capabilities.

Besides genetic algorithms and genetic programming, there are also other evolutionary approaches that can be applied in a global induction. In [82], a differential evolution-based system for finding a near-optimal axis-parallel decision tree is introduced. It uses linear real-valued chromosomes for encoding the internal nodes of a classification tree. The size of the evolved trees is fixed and is related to the number of features in the training data. The reclassification accuracy is utilized as a measure of the fitness, so post-evolutionary processing is necessary, and this encompasses error-based pruning [22].

Gene expression programming (GEP) [83] is another interesting evolutionary technique proposed by C. Ferreira and derived from genetic programming and genetic algorithms. It uses a fixed-sized linear genome composed of a head and tail, which are linked but interpreted differently. The author describes two applications aimed at a simple induction of a decision tree with nominal attributes and more advanced variant called EDT-RNC (*Evolvable Decision Trees with Random Numerical Constants*) for handling nominal and numeric attributes. Other authors have tried to apply this technique to generate univariate classification trees [84]; however, their system called GEPDT, is rather limited because it exploits only equality nominal tests. In [85], two GEP-based ensemble classifiers are proposed and experimentally validated.

Recently, even a coprocessor for hardware-aided full decision tree induction using an evolutionary approach has been presented [86]. The authors develop the FPGA-based acceleration of the fitness evaluation.

Finally, applying evolutionary computations in the induction of decision trees has been an active and successful research area over the last twenty years. Even surveys concerning on subject have been published in the most prestigious journals ([87] or [88]).

References

1. Chen C, Zhang C (2014) Inf Sci 275:314–347
2. Wu P, Cheng C, Kaddi C, Venugopalan J, Hoffman R, Wang M (2017) IEEE Trans Biomed Eng 64(2):263–273
3. Zhong R, Huang G, Lan S, Dai Q, Chen X, Zhang T (2015) Int J Prod Econ 165:260–272
4. Gungor V, Sahin D, Kocak T, Ergut S, Buccella C, Cecati C, Hancke G (2013) IEEE Trans Ind Inform 9(1):28–42
5. Emani C, Cullot N, Nicolle C (2015) Comput Sci Rev 17:70–81
6. Gupta U, Gupta A (2016) J Int Bus Res Mark 1(3):50–56
7. Fayyad U, Uthurusamy R (2002) Commun ACM 45(8):28–31
8. Vassiliadis P (2009) Int J Data Warehous Min 5(3):1–27
9. Wu X, Zhu X, Wu G, Ding W (2014) IEEE Trans Knowl Data Eng 26(1):97–107
10. Fayyad U, Piatetsky-Shapiro G, Smyth P, Uthurusamy R (1996) Advances in knowledge discovery and data mining. AAAI Press
11. Friedman J, Hastie T, Tibshirani R (2001) The elements of statistical learning. Springer, Berlin
12. Duda O, Heart P, Stork D (2001) Pattern classification. 2nd edn. Wiley, New York
13. Mitchell T (1997) Machine learning. McGraw-Hill, New York
14. Hand D, Mannila H, Smyth P (2001) Principles of data mining. The MIT Press, Cambridge
15. McGarry K (2005) Knowl Eng Rev 20(1):39–61
16. Domingos P (2012) Commun ACM 55(10):78–87
17. Liu H, Hussain F, Tan C, Dash M (2002) Data Min Knowl Discov 6(4):393–423
18. Kotsiantis S (2013) Artif Intell Rev 39:261–283
19. Rokach L, Maimon O (2014) Data mining with decision trees: theory and applications, 2nd edn. World Scientific
20. Polikar R (2006) IEEE Circuits Syst Mag 6(3):21–45
21. Krizhevsky A, Sutskever I, Hinton G (2012) Imagenet classification with deep convolutional neural networks. In: Advances in neural information processing systems, pp 1097–1105
22. Quinlan J (1993) C4.5: programs for machine learning. Morgan Kaufmann, San Francisco
23. Murthy S (1998) Data Min Knowl Discov 2:345–389
24. Utgoff P (1989) Connect Sci 1(4):377–391
25. Llora X, Wilson S (2004) Mixed decision trees: minimizing knowledge representation bias in LCS. In: Proceedings of GECCO'04. Lecture notes in computer science, vol 3103, pp 797–809
26. Yildiz O, Alpaydin E (2001) IEEE Trans Neural Netw 12(6):1539–1546
27. Loh W-Y (2014) Int Stat Rev 82(3):329–348
28. Quinlan J (1986) Mach Learn 1(1):81–106
29. Kass G (1980) Appl Stat 29(2):119–127
30. Brodley C, Utgoff P (1995) Mach Learn 19(1):45–77
31. Breiman L, Friedman J, Olshen R, Stone C (1984) Classification and regression trees. Wadsworth and Brooks, Monterey
32. Murthy S, Kasif S, Salzberg S (1994) J Artif Intell Res 2:1–33
33. Gama J, Brazdil P (1999) Intel Data Anal 3(1):1–22
34. Quinlan J (1992) Learning with continuous classes. In: Proceedings of AI'92, pp 343–348
35. Torgo L (1997) Functional models for regression tree leaves. In: proceedings of ICML'97. Morgan Kaufmann, pp 385–393
36. Hayfil L, Rivest R (1976) Inf Process Lett 5(1):15–17
37. Brassard G, Bratley P (1996) Fundamentals of algorithmics. Prentice Hall
38. Rokach L, Maimon O (2005) IEEE Trans SMC C 35(4):476–487
39. Esposito F, Malerba D, Semeraro G (1997) IEEE Trans Pattern Anal Mach Intell 19(5):476–491
40. Quinlan J (1987) Int J Man Mach Stud 27:221–234
41. Bobrowski L (1996) Piecewise-linear classifiers, formal neurons and separability of the learning sets. In: Proceedings of 13 ICPR. IEEE computer society press, pp 224–228
42. Czajkowski M, Kretowski M (2014) Inf Sci 288:153–173

43. Shah S, Sastry P (1999) IEEE Trans SMC C 29(4):494–505
44. Vogel D, Asparouhov O, Scheffer T (2007) Scalable look-ahead linear regression trees. In: Proceedings of KDD'07. ACM Press, New York, pp 757–764
45. Wang Y, Xia S, Wu J (2017) Knowl-Based Syst 120:34–42
46. Kozak J (2019) Decision tree and ensemble learning based on ant colony optimization. Springer, Berlin
47. Freitas A (2002) Data mining and knowledge discovery with evolutionary algorithms. Springer, Berlin Heidelberg
48. Chai B, Huang T, Zhuang X, Zhao Y, Sklansky J (1996) Pattern Recognit 29(11):1905–1917
49. Cantu-Paz E, Kamath C (2003) IEEE Trans Evol Comput 7(1):54–68
50. Ng S, Leung K (2005) Induction of linear decision trees with real-coded genetic algorithms and k-D trees. In: Proceedings of IDEAL'05. Lecture notes in compter science, vol 3578, pp 264–271
51. Tan P, Dowe D (2004) MML inference of oblique decision trees. In: Proceedings of AJCAI'04. Lecture notes in computer science, vol 3339, pp 1082–1088
52. Kretowski M (2004) An evolutionary algorithm for oblique decision tree induction. In: Proceedings of ICAISC'04. Lecture notes in artificial intelligence, vol 3070, pp 432–437
53. Vilalta R, Drissi Y (2002) Artif Intell Rev 18(2):77–95
54. Barros R, Basgalupp M, Freitas A, Carvalho A (2014) IEEE Trans Evol Comput 18(6):873–892
55. Karabadji N, Seidi H, Bousetouane F, Dhifi W, Aridhi S (2017) Knowl-Based Syst 119:166–177
56. Frank E, Hall M, Witten I (2016) The WEKA workbench. Online appendix for "data mining: practical machine learning tools and techniques", 4th edn. Morgan Kaufmann, San Francisco
57. Koza J (1992) Genetic programming: on the programming of computers by means of natural selection. MIT Press, Cambridge
58. Koza J (1991) Concept formation and decision tree induction using genetic programming paradigm, In: Proceedings of PPSN 1. Lecture notes in computer science, vol 496, pp 124–128
59. Nikolaev N, Slavov V (1998) Intell Data Anal 2:31–44
60. Tanigawa T, Zhao Q (2000) A study on efficient generation of decision tree using genetic programming. In: Proceedigns of GECCO'00, pp 1047–1052
61. Bot M, Langdon W (2000) Application of genetic programming to induction of linear classification trees. In: EuroGP 2000. Lecture notes in computer science, vol 1802, pp 247–258
62. Folino G, Pizzuti C, Spezzano G (1999) A cellular genetic programming approach to classification. In: Proceedings of GECCO'99, Morgan Kaufmann, pp 1015–1020
63. Folino G, Pizzuti C, Spezzano G (2000) Genetic programming and simulated annealing: a hybrid method to evolve decision trees. In: EuroGP'00, Lecture notes in computer science, vol 1802, pp 294–303
64. Folino G, Pizzuti C, Spezzano G (2002) Improving induction decision trees with parallel genetic programming. In: Proceedings of EUROMICROPDP'02, IEEE Press, pp 181–187
65. Kuo C, Hong T, Chen C (2007) Soft Comput 11:1165–1172
66. Saremi M, Yaghmaee F (2018) Comput Intell 34:495–514
67. Papagelis A, Kalles D (2001) Breeding decision trees using evolutionary techniques. In: Proceedings of ICML'01. Morgan Kaufmann, pp 393–400
68. Kalles D, Papagelis A (2010) Soft Comput 14(9):973–993
69. Fu Z, Golden B, Lele S, Raghavan S, Wasil E (2003) INFORMS J Comput 15(1):3–22
70. Fu Z, Golden B, Lele S, Raghavan S, Wasil E (2003) Oper Res 51(6):894–907
71. Sorensen K, Janssens G (2003) Eur J Oper Res 151:253–264
72. Llora X, Garrell J (2001) Evolution of decision trees. In: Proceedings of CCAI'01. ACIA Press, pp 115–122
73. Cha S, Tappert C (2009) J Pattern Recognit Res 4(1):1–13
74. Fan G, Gray JB (2005) J Comput Graph Stat 14(1):206–218
75. Schwarz G (1978) Ann Stat 6:461–464
76. Gray J, Fan G (2008) Comput Stat Data Anal 52(3):1362–1372
77. Hazan A, Ramirez R, Maestre E, Perez A, Pertusa A. (2006) In: Applications of evolutionary computing. Lecture notes in computer science, vol 3907, pp 676–687

78. Barros R, Ruiz D, Basgalupp M (2011) Inf Sci 181:954–971
79. Potgieter G, Engelbrecht A (2008) Expert Syst Appl 35:1513–1532
80. Potgieter G, Engelbrecht A (2007) Appl Math Comput 186(2):1441–1466
81. Sprogar M (2015) Genet Program Evolvable Mach 16:499
82. Rivera-Lopez R, Canul-Reich J (2018) IEEE Access 6:5548–5563
83. Ferreira C (2006) Gene expression programming: mathematical modeling by an artificial intelligence. Springer, Berlin
84. Wang W, Li Q, Han S, Lin H (2006) A preliminary study on constructing decision tree with gene expression programming. In: Proceedings of ICICIC'06. IEEE computer society, vol 1, pp 222–225
85. Jedrzejowicz J, Jedrzejowicz P (2011) Expert Syst Appl 38(9):10932–39
86. Vukobratovic B, Struharik R (2016) Microprocess Microsyst 45B:253–269
87. Barros R, Basgalupp M, Carvalho A, Freitas A (2012) IEEE Trans SMC C 42(3):291–312
88. Podgorelec V, Sprogar M, Pohorec S (2013) WIREs Data Min Knowl Discov 3:63–82

Chapter 3
Parallel and Distributed Computation

Big Data mining is an excellent example of an algorithmic challenge that, without a doubt requires a huge amount of computational resources. Nowadays, computers are quite good for typical document editing or Internet exploration, but they are not sufficient for highly demanding calculations. Moreover, it is well-known that improving the computation power of stand-alone machines by increasing the clock frequency in processors is limited. We have to admit that the famous Moore's law [1] concerning a doubling of a number of transistors per chip every eighteen months is not valid anymore because of reaching physical and technological limits. Currently, the most straightforward way to improve computational performance is to use multiple processing units to solve the same problem.

The motivations behind parallel processing are rather straightforward; first of all, it is the computation time that matters. Almost instant reactions based on vital simulations and analyses are indispensable in very competitive businesses and modern industry. Because potentially useful and vast data can be acquired easily, the mining of bigger and bigger datasets is highly appreciated, and without the intensive application of concurrency, it is obviously impossible. Moreover, data, as well as computational resources, are becoming naturally scattered, so distributed processing is not only useful but necessary.

When considering parallel or distributed implementations, we cannot forget that if not realized even the best ideas remain only dreams. Here, a key element is access to suitable hardware resources. It is clear that with extremely expensive super-computers, the results can be spectacular, but in practice, we have to be more interested in reasonable yet costly architectures and be guided by the benefit-cost ratio. Hopefully, nowadays, efficient parallel and distributed solutions can be developed with the commodity systems that come at a moderate cost, and in this book, I concentrate on them.

There are two main directions in which a classical personal computer with a uniprocessor (so-called SISD architecture of Flynn's taxonomy [2]: single instruction stream, single data stream) can evolve. The first broad category of parallel architectures encompasses MIMD (multiple instruction stream, multiple data stream) solutions, which can be created from many processing units executing diverse instructions

© Springer Nature Switzerland AG 2019

M. Kretowski, *Evolutionary Decision Trees in Large-Scale Data Mining*,
Studies in Big Data 59, https://doi.org/10.1007/978-3-030-21851-5_3

with different data. The second direction focuses on the SIMD (single instruction stream, multiple data stream) approach, where the processing units execute the same instructions on different data. Currently, the graphical processing units used for general processing [3] and specialized computing accelerator cards [4] are good representatives of this approach, and the devices of these types are intensively developed.

Concerning the MIMD systems, based on their memory structure and communication mechanism, two of the most important subcategories are distinguished as follows [5]:

- GMSV (global memory, shared variables)—known also as the shared-memory approach (multiprocessor). Processors (or cores) communicate through the single memory address space, and the memory is physically shared by all processing units. A multicore processor and multiprocessor machine are typical examples of this approach. In a multiprocessor machine, if all the processors have uniform access to all the resources (e.g., memory or input/output devices), such a computer is called a symmetric multiprocessor (SMP) machine. When concerning memory access in a multiprocessor machine, two architectures can be distinguished: uniform memory access (UMA) with the same delay of memory accessing for any processor, and non-uniform memory access (NUMA), where the access to the local memory is faster.
- DMMP (distributed memory, message passing)—called the distributed-memory approach (multicomputer). Message passing is used by independent computers to ensure inter-processor communication. The memory is physically distributed between the machines. Computer clusters and grids [6] are representative examples of this approach.

In Fig. 3.1, three types of MIMD memory architecture are presented. It should be noted that the global memory address space provides a user-friendly programming perspective, and memory access is fast and uniform because of the proximity of the memory to the processors. On the other hand, distributed memory is better able to scale with the number of processors, but programmers are responsible for many of the communication details. The hybrid approach tries to balance the advantages and disadvantages of these two architectures.

The rest of this chapter is organized as follows: In the next section, the major ideas of parallel algorithm design are discussed, and then, a performance evaluation of parallel implementations is shortly described. In the following sections, selected parallel and distributed approaches are briefly introduced. Only approaches that will be afterward applied to speed-up the global induction of decision trees are included.

3.1 Parallel Algorithm Design

The design of an efficient sequential algorithm that solves a non-trivial problem is typically a challenging work, but in the case of a parallel algorithm, the level of difficulty raises significantly. Apart from these standard limitations and difficulties, one has to deal with completely new issues specific only to both a parallel hardware

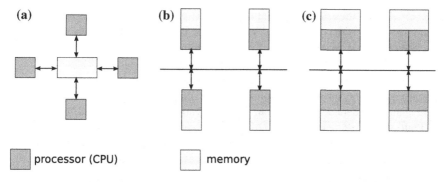

Fig. 3.1 Three types of MIMD memory architectures: **a** shared, **b** distributed, and **c** hybrid distributed-shared

and programming model. For example, it is usually necessary to have communication between concurrent tasks and their synchronization. There are no universal rules on how to design a parallel program or how to convert a sequential algorithm into its concurrent version, but most scholars agree that certain elements are crucial for the process [7, 8]. Below, the main elements are briefly discussed.

3.1.1 Partitioning

As a first step in algorithm design, both the necessary computation and data to be operated on have to be identified and understood in detail. This should enable partitioning them into possible small chunks that can be treated as abstract building blocks. In parallel processing, tasks are used as basic realization elements, and they are described as programmer-defined units of computation that may be executed in parallel. Such low-level tasks are regarded as indivisible, and their size can be arbitrary although the equal size of the tasks can be convenient. The tasks can be obviously merged, grouped, and run together, but these actions are decided later in the aggregation and mapping stage.

There are two major approaches to a problem decomposition: a domain decomposition, which comes from the processed data, and functional decomposition, which concentrates on the work that has to be done. Most of the algorithms are data intensive because they process input data, create intermediate data, or produce output data. Because effectiveness is the ultimate goal of a parallel implementation, it generally makes the most sense to focus on the largest piece of data or the most frequently used. Each parallel task can work on a portion of the data, and usually, there are different ways to partition the data depending on the realized activity. For example, in a typical data mining algorithm, an input dataset can be decomposed vertically (by instances) or horizontally (by features), or the decomposition can exploit two dimensions. The partitioning can also be cyclic or block like. When considering the different possibilities of the parallel solving of a problem or analyzing the variants of an algorithm, it may be convenient to use the degree of concurrency concept. It describes the (maximal) number of uniform tasks that can be realized at the same moment. In the aforementioned dataset decomposition example, because the number of instances is typically much higher than the number of features, the degree of concurrency of the vertical partitioning is much bigger.

In contrast to data decomposition, the second approach tries to partition the workload into separate activities, which often correspond to specific functions. Each parallel task performs a fraction of the total computation. In the simplest scenario, the tasks operate on disjointed data, and communication is not necessary. In more frequent situations, the data processed by various tasks overlaps, so data replication or synchronized communication can be required. In the most severe overlapping, this can forcibly rule out the functional decomposition. In general, a degree of concurrency, which is possible to achieve, is lower for this type of partitioning. However, it is advisable to consider many variants of both decomposition types because the most suitable solution is strongly related to the problem being solved. It is also possible that a hybrid algorithm can result in the best performance.

3.1.2 Communication

Communication between tasks is an indispensable part of any parallel algorithm. Only for a small group of problems will communication requirements be reduced to a minimum, which obviously eases the algorithm design. Such problems are called embarrassingly parallel [9]. Notwithstanding, most parallel algorithms require tasks to share data with each other, and having a cost related with communication and synchronization can be important. Depending on the level of granularity (understood as the ratio of computation to communication), loosely coupled and tightly coupled problems can be distinguished. The problems of the second type are perceived as the most difficult to parallelize.

Depending on the parallel system's architecture, communication design can be described in two steps. First, the communication channels are identified, and, then, an actual way to communicate is decided. The channel structure is especially important for distributed architectures, where sending and receiving messages are associated with noticeable costs. Computational resources may be needed to package and transmit data. If a frequent synchronisation is necessary, it can cause some tasks to wait instead of working, or the network bandwidth may saturate. Two concepts to characterize the transmission process (e.g., transmitting a message between two processors or transferring data between cache levels and processors) [9] are noted as follows:

- Latency—the time needed to set up and start the transmission, independent of the message size;
- Bandwidth—after a transmission has started, the number of bytes per second that can go through the channel.

In other words, latency is the time it takes to transfer a minimal message between two tasks, whereas bandwidth is the amount of data that can be transferred per unit of time. Latency can dominate a communication's overhead when many small messages are sent.

An important design decision that is devoted to the synchronous or asynchronous modes of communication. In the synchronous mode, tasks that are involved in the data transfer are excluded from other activities. They can realize subsequent instructions only when a confirmation of the transfer end is delivered. This mode of communication is also referred to as blocking communication. In the case of an asynchronous mode, tasks are not strictly linked during the transfer. A source task simply reports that a message is ready to be sent and proceeds to further instructions. The message will be received by a recipient task when the task is ready for this. As expected, such a mode of communication is called non-blocking. Asynchronous communication is generally more difficult to implement and less predictable, but on the other hand, it can be very efficient because it enables overlapping the computation with communication.

The scope of communication is another crucial aspect. Local communication, which encompasses only a small number of tasks, is the most desired communication pattern. For task pairs, it is the simplest to define the connecting channels (point-to-point communication) and to implement adequate send/receive functions. In a certain situation, global communication is unavoidable, and it can involve all tasks. Typically, this type of communication is supported by collective operations that enable the expression of complex data transfers using simple semantics. Broadcast (one-to-all) and gather (all-to-one) are good examples of collective operations that drive exchanges of data between a single task and all other tasks in a communication group. Well-optimized collective communication routines are available virtually in all parallel programming libraries, and their use not only improves the code's readability, but can also enhance the algorithm's performance.

3.1.3 Agglomeration and Mapping

When the possible degrees of concurrency are thoroughly studied and needful communication patterns with their costs are known, it is time to match them with available parallel system's architecture. Its limitations and preferences should be confronted with the needs and expectations of the considered parallel algorithm. Agglomeration is an activity that leads to the task grouping or merging. In certain situations, by increasing the algorithm's granularity, it could reduce or even eliminate the impact of communication on the algorithm performance.

Reducing the time spent on communication, because it usually may stop computation, is one of the most obvious objectives of the agglomeration. For example, if a group of tasks that intensively communicates with each other can be merged, this increases the data locality and eliminates communication. Similarly, if we have a group of tasks that regularly sends messages to the other group, it could be profitable to combine the sending group into one task and the receiving group into the second one.

Another technique to reduce the communication time deals with data or computation replication. It is possible that in a certain condition, replication can be less costly than sending additional messages. Here, a reduction in communication time is not the only goal of agglomeration, and any aggregation activities make sense if the overall performance is improved.

In the final step, which is called mapping, the primitive or aggregated tasks are assigned to processors for execution. Typically, the shortest execution time can be obtained with the most effective use of all the available resources. Apart from the previously discussed communication overhead, the processor's idle time is the second obstacle. There are two sources of processor idleness. In the first situation, a processor can just wait for a (new) task, for example, after completing the previous one while the other processors are still working. In the second situation, a processor can wait for the data necessary to realize the task, for example, when the expected data have not yet been produced by another task or when the processors have to compete for access to the data. Load balancing algorithms [10] try to equalize the workload distribution between processors, and they are sufficient for the first situation but not for the second one. Here, a very attentive analysis of time dependencies and data access patterns usually is necessary for identifying a good mapping. There are two categories of mapping:

- Static mapping—it decides the task distribution to processors before start up and the mapping cannot be changed during the algorithm's execution. These methods can be successful only if the task execution can be precisely characterized in advance for a given parallel computer;
- Dynamic mapping—it distributes the tasks among the processors based on the current processor load status. This type of mapping is indispensable when the task workloads are uneven or unpredictable. It may also be preferable for heterogenous parallel machines or when the computing resources have to be shared with external systems. Dynamic load balancing can be realised as centralized or distributed.

3.2   Performance Evaluation of Parallel Implementations

From a user perspective, on any data mining system, the execution times observed for a given training dataset on available hardware are the most straightforward measures of the applicability of different algorithms or methods. For research purposes, such a simple rating is usually not enough, and more elaborate coefficients are analyzed. Speedup and efficiency are two basic measures used for the evaluation of parallel implementation performance [9]. *Speedup S* quantifies how much faster a parallel algorithm is than a corresponding sequential algorithm:

$$S(p) = \frac{T_s}{T(p)}, \tag{3.1}$$

where p is the number of computing units (processors or cores), $T(p)$ is the total parallel execution time of the algorithm on p units, and T_s is the execution time of the sequential version. If T_s corresponds to the best possible sequential implementation, the obtained speedup is called *absolute*. Whereas a *relative* speedup is measured, when the same (parallel) algorithm is launched on one and p processors ($T_s = T(1)$).

In a perfect parallelization scenario, the obtained speedup is equal to the number of processors ($S(p) = p$), and this is called a *linear* speedup. However, in typical situations, a linear speedup is rarely observed because there is usually a certain overhead (e.g., because of synchronization, communication, or data transfer), and also, not all the components of the algorithm may be parallelized. One of the first and well-known models that tries to describe a relation between speedup and the number of processors is *Amdahl's law* [11]. This law takes the assumption that a certain part of the algorithm has to be realized in a sequential manner and could not be parallelized. The execution time of this part can be expressed as $\zeta * T_s$, where $\zeta \in [0, 1]$ is a non-parallelizable computation fraction of the whole algorithm. Thus, the total execution time on p processors is equal to:

$$T(p) = \zeta * T_s + (1 - \zeta) * \frac{T_s}{p} \tag{3.2}$$

and as a result, speedup can be represented by:

$$S(p) = \frac{p}{\zeta(p - 1) + 1}. \tag{3.3}$$

It is clear that the non-parallelizable fraction has a crucial impact on the acceleration that can be achieved.

It should be noted that if the non-parallelizable portion of a program takes half of the sequential runtime ($\zeta = 0.5$), it is not possible to get a speedup above two, no matter how many processors are available. For $\zeta = 0.1$, the speedup limit is equal to ten, and to get a speedup exceeding 100, no more than one percent of the code could be realized sequentially. In practice, the situation is even more complicated because communication overhead and load unbalance have to be taken into account. In Fig. 3.2, an example of a more realistic situation is presented.

Also, a speedup larger than the number of processors used (a *superlinear* speedup) can be observed in specific conditions. For example, this can be related to the better use of memory caches in the case of data parallelism or possible simplifications of the processing when a parallel version of the algorithm is implemented.

The highest possible speedup is often the ultimate goal of parallelization efforts; however, the speedup is closely related to the number of processors. Usually, better hardware could improve the obtained result, but a certain part of the resources used could be wasted. In such a situation, the second performance measure, namely effi-

Fig. 3.2 An example of a realistic speedup observation

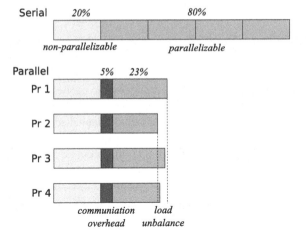

ciency, could be helpful. *Efficiency E* is defined as the ratio of the speedup to the number of processors:

$$E(p) = \frac{S(p)}{p}.$$ (3.4)

This can be treated as an estimate of the fraction of time for which processors are usefully utilized. For a perfect, theoretical parallelization, the obtained efficiency is equal to one1 (sometimes given as 100 percent), but in practice, the observed efficiency is noticeably lower and often decreases when additional processors are included. The ability of a parallel program to take advantage of an increasing number of processors is usually measured by *scalability*, and a scalable program can maintain efficiency at the same level with a growing number of processing units.

3.3 Distributed Memory Programming Through Message Passing

In distributed systems where processes do not have access to a shared memory, the communication between parallel tasks is accomplished by the explicit sending and receiving of messages. This is typically realized using a Message Passing Interface (MPI) [12], which is a specification for the developers and users of message passing libraries. Within the message-passing parallel programming model, data are moved from one process to another process through cooperative operations on each process. MPI represents explicit parallelism, meaning that a programmer is responsible for correctly identifying parallelism and implementing parallel algorithms using MPI constructs. This requires not only standard programming skills, but also an understanding of distributed and/or parallel architecture and the programming model. Because of the non-deterministic character of the parallel program execution, complex time dependencies can be observed, which can lead to unexpected behaviours (a race problem). Generally, the MPI programming is perceived as one of the most difficult programming activities.

The MPI routines can be assigned to the following major categories [9]:

- Process management—setting up a parallel execution environment and querying it;
- Point-to-point communication—interactions between two processes, mostly sending and receiving data. The standard communication is blocking,[1] but non-blocking versions of the routines are also available;
- Collective calls—interactions among many processors (encompassing all the processors or a specified subset).

In the first category, at least a few routines need to be mentioned. MPI_Init initializes the MPI execution environment. This function must be called only once and before any other MPI functions. On the other end, MPI_Finalize() terminates the MPI execution environment. Because in the MPI programs all processes execute the same program, they have to be able to recognize collaborating resources and identify themselves. This allows for carrying out different actions in specific processors. The MPI_Comm_size(comm, &size) routine returns the total number of MPI processes in the specified communicator (MPI_COMM_WORLD is typically used as comm for all processes), whereas MPI_Comm_rank(comm, &rank) returns the rank (sometimes called a task ID) of the calling MPI process within the specified communicator. It should be mentioned that when the MPI program is launched (with mpiexec), the user declares the number of collaborating processes that should be created.

For point-to-point communication, the most basic blocking pair of send and receive routines is as follows:

```
MPI_Send(&buf, count, datatype, destination, tag, comm),
MPI_Recv(&buf, count, datatype, source, tag, comm, &status).
```

The message buffer is declared by three parameters: (buf references a memory area, where the data are stored, count describes how many elements should be transferred, and datatype defines a type of elements. Both elementary (e.g., MPI_CHAR and MPI_DOUBLE correspond to char and double) or more complex data types, defined by a user like structures, can be used. Various message types can be introduced and identified by user-defined tags (tag parameters). The target or source processes are specified by the comm communicator and destination or source ranks. Receiving more elements that are declared by count results in an error. The status parameter can contain additional information, for example, on how much data were actually received.

---

[1]Blocking communication—in a certain simplification, the first processor executes the send instruction, and then, it waits (it is temporarily blocked) until the second processor acknowledges that the data were successfully received.

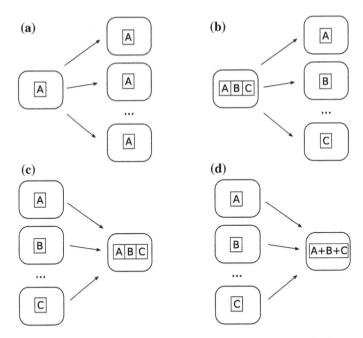

Fig. 3.3 Main collective operations: **a** broadcast, **b** scatter, **c** gather, and **d** reduction

Collective communication involves the participation of all processes in a communicator. There are different types of collective operations (a few popular operations can be seen in Fig. 3.3) that are typically grouped in three categories:

- Synchronization—the processes wait until all members of the group have reached a point in the code, and then, they are free to proceed; barrier synchronization is realized by `MPI_Barrier(comm)` and can be helpful for measuring performance;
- data movements—encompass the variants of broadcasting, scattering, gathering, and all-to-all transferring;
- collective computation—the values of a variable from the processors are combined; reduction and scan are good examples.

The simplest data movement operation is broadcast, which sends the same data from one processor (the root) to all processors in a communicator. It might appear as though `MPI_Bcast` could be just a simple wrapper around `MPI_Send` and `MPI_Recv`, but its efficient implementation utilizes a tree broadcast algorithm for the best network utilization. Sending distinct data for every processor in a communicator can be realized by `MPI_Scatter`, whereas an inverse operation can be offered by `MPI_Gather`. More complex communication patterns can also be used, and finally, all-to-all transfers are enabled (`MPI_Alltoall`).

In many practical applications, broadcasting is combined with a reduction. `MPI_Reduce` performs a global calculation by aggregating the values of a variable using a commutative binary operator `op`:

```
MPI_Reduce(&sbuf, &rbuf, count, datatype, op, root, comm).
```

Both predefined (e.g., simple sum—MPI_SUM) and user-supplied operators can be used. The `root` rank identifies a processor where the reduced value is placed.

3.4 Shared Memory Programming

Communication through message passing is not necessary when the processes can have direct access to the same physical memory with a single address space. Today, almost all machines, starting from smartphones and laptops up to computing stations are equipped with multicore processors with shared memory. This enables very fast data exchange between tasks, and the programming arsenal can be simplified. In shared memory programming [9], a crucial position is occupied by threads.

A thread is an executable, partially independent component within the process. It is the smallest unit of processing that can be scheduled by an operating system. Multiple threads can exist within one process, and they can execute concurrently. Threads are dynamic, and their number can change during runtime. When a program starts, there is only one thread (master thread), and other threads can be created by thread spawning and can form a thread team. The master thread can wait for the completion of the thread team's work. This type of parallelism is called a fork-join model and schematically is represented in Fig. 3.4.

Although communication through shared variables facilitates programming, it can also cause problems. For example, a race condition may occur when at least one thread tries to modify a location in shared memory that is supposed to be accessed by other threads. Because the access order is not known, the resulting behavior cannot be predicted. Only the explicit coordination of memory access avoids the problem.

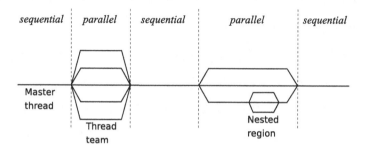

Fig. 3.4 Fork-join parallelism

This can be realized by a barrier synchronization[2] or by memory locks. Obviously, it can slow down the program execution, or it can even lead to a deadlock when threads are blocked mutually. Generally, it is advisable to minimize data sharing.

Currently, OpenMP (Open Multi-Processing) [13, 14] is a dominant solution for implementing multi-threaded applications meant for shared memory machines. There are three primary components in this API: compiler directives (with clauses), runtime library routines, and environment variables.

Parallel regions are marked by the compiler directives. In C/C++, these directives take the form of `pragma` statements, and their common format is as follows:

```
#pragma omp directive_name [clause_list]
```

Each directive is applied to at most one succeeding statement that must be a structured block (such a pair is called a construct). The fundamental directive is `parallel`, and it enables the creation of a parallel region with a given number of threads, like in the following example, where a parallel region with ten threads is created:

```
#pragma omp parallel num_threads(10)
```

It is possible to introduce a conditional parallelization with the `if(scalar _expression)` clause. If such a clause is present, it must evaluate to non-zero for a team of threads to be created. Otherwise, the region is executed serially by the master thread.

In OpenMP programs, variables declared before a parallel block can be treated as shared or private in this block. Shared variables are declared by the `shared(variable_list)` clause and are visible by all threads in a team, whereas the `private(variable_list)` clause restricts access only to one thread each. From a performance perspective, it is clearly advisable to avoid shared data.

Private variables are not initialized by default before entering a parallel block. By using the `firstprivate(variable)` clause, all copies are initialized with the corresponding value from the master thread. Similarly, the `lastprivate(variable)` clause enables updating the variables when going out of a region. The `default(mode)` clause controls the default data-sharing attributes of the variables. It is advisable to use `default(none)` to be not surprised by default sharing.

One of the most common OpenMP applications is the `for` loop parallelization. Typically, the `for` shortcut can be used, and iterations are realized concurrently by threads (but their order is not preserved):

```
#pragma omp parallel for
```

---

[2] A barrier mechanism concerns all threads and enforces faster threads to wait for the rest at a chosen location of a program's code.

Iterations have be independent to be safely executed in any order without having loop-carried dependencies. It is possible to influence how the iterations are grouped and executed by the `schedule(scheduling_class [,chunk_size])` clause. `Chunk_size` specifies the number of iterations assigned to one thread at a time. There are three choices for loop scheduling:

- `static`—scheduling is predefined at the compilation time, and iteration chunks are assigned to threads in a round-robin fashion; it has the lowest overhead and is predictable;
- `dynamic`—selection is made at runtime, and groups are assigned to the next available thread;
- `guided`—special case of dynamic that attempts to reduce overhead.

In loops, one common pattern is reduction, where through iterations, values are accumulated into a single value. Thanks to the `reduction(op:var_list)` clause, each thread can have local copies of (shared) variables from `var_list`, which are modified independently with no risk of facing race conditions. And finally, local copies are only reduced using the `op` operator into a single value and combining it with the initial value.

For the parallelization of non-iterative tasks (e.g., different functions), another clause can be used:

```
#pragma omp parallel sections
```

Each section (preceded by `#pragma omp section`) is executed by one thread from the team.

One very positive feature of OpenMP is how easy it is to start using it because it is possible to begin just with a sequential version of a code and incrementally transform it into a parallel one. By contrast, in the case of MPI-based code re-engineering, it is an all-or-nothing affair [9].

A few examples of useful runtime library functions (C style) are as follows:

- `void omp_set_num_threads(int nThreads)`—sets the number of threads in a team;
- `int omp_get_thread_num()`—returns a thread identifier;
- `int omp_get_num_procs()`—returns the number of available processors;
- `int omp_get_num_threads()`—returns the number of threads in a current team.

The execution of the OpenMP application can also be altered by using environmental variables. For example, the number of threads can be specified by `OMP_NUM_THREADS num`.

Synchronization is used to impose order constraints and protect access to shared data. OpenMP provides a set of synchronization directives. The simplest synchronization directive is `barrier`—each thread waits until all threads from a team arrive. Barrier-like synchronization is implicitly performed at the end of a `for` work-sharing construct (except when a `nowait` clause is added). The `critical` clause preceding a block of code can be used when a critical section is necessary. It provides a mutual exclusion among the threads executing this block of code. Only one thread at a time enters a critical region. The `atomic` clause is a simplified version of a critical section, where only the read/update of a single memory location is guarded. In most machines, these atomic operations are much more efficient because hardware support. Even in a parallel region, it is sometimes useful to have only one thread executing certain lines of the code, for example, corresponding to the input/output operations. The `master` directive specifies a code region that is to be executed only by the master thread of the team, and all other threads skip this section. The `single` directive is very similar because the marked section is executed just by one thread, but it is not known which one.

MPI and OpenMP should not be treated as competitors because they are mainly designed for different architectures. Furthermore, they can collaborate perfectly in a hybrid hardware environment of clustered multi/many cores machines. Threads can perform computationally intensive tasks using local, on-node data, whereas the communication between processes on different nodes occurs over the network by using message passing.

3.5 GPGPU and CUDA

Graphical cards have evolved from specialized hardware that could process and render visual data to becoming real parallel mini-supercomputers. When researchers and engineers realized the computational power of graphics processing units (GPUs) in solving inherently parallel generic problems, almost a new discipline emerged: general-purpose computation on GPUs (GPGPU) [3]. Now, a GPU is often considered a co-processor that can execute thousands of threads in parallel to handle the tasks traditionally performed by a CPU. Various approaches and frameworks have arisen for facilitating the use of GPU-based computation resources. OpenCL (Open Computing Language) [15] is one of the most general and flexible heterogenous computing frameworks and was developed by the Khronos Group, whereas Compute Unified Device Architecture (CUDA) [16] was created by the NVIDIA Corporation and seems to be the most popular and efficient vendor-specific approach. It is worth mentioning that according to A. Cano, CUDA is more frequently used than OpenCL in the data mining research community [17].

In Fig. 3.5, a typical CUDA GPU architecture is schematically presented. The GPU engine is a scalable array of streaming multiprocessors (SMs). Each SM has multiple simple streaming processors (called CUDA cores) but only one instruction unit. As a result, a group of processors runs exactly the same set of instructions at any given time within a single SM. This means that CUDA employs the classical SIMD (Single-Instruction Multiple-Data) parallelism from Flynn's taxonomy [8].

The GPU's memory is both hierarchical and heterogenous [18]. A few distinct memory types can be distinguished with different scopes, lifetimes, and caching

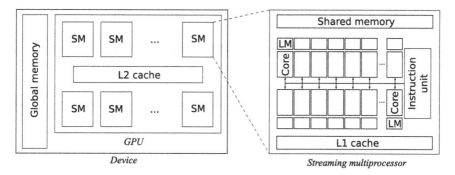

Fig. 3.5 Typical CUDA GPU architecture (LM stands for local memory)

behaviors. The global memory has the largest capacity, and all SMs have access to all the global memory. However, this memory has the highest latency. Because global memory is high bandwidth, it permits high-efficiency access patterns by using data coalescing. This requires algorithms that use consecutive threads to access consecutive data memory positions, minimizing the number of memory transactions. Faster, on-chip memory is assigned to each SM. All cores inside the same SM share some limited memory space, and they have their own small local memories. The fastest access is attributed to caches and registers.

In the CUDA programming model, a GPU is called a device, whereas a CPU is called a host. When the host wants to delegate any job to the device, it calls a kernel, which is a special function run on the device. The programmer only specifies an intended two-level structure of threads that execute the same kernel code in parallel. The framework is responsible of launching the threads and mapping them on SMs and cores. The first level of the hierarchical structure is known as a grid, which is composed of thread blocks, and the second level organizes these threads into blocks. Each thread is associated with an index, which the recovery of the thread position in the structure and the located data to be processed or to control the algorithmic flow. Each thread block is mapped to a single SM, and then, the threads from the block are mapped to the cores.

When any kernel is called, the grid of blocks must be decided on in advance:

```
KernelName<<<dimGrid, dimBlock>>>(Par1, Par2, ....);
```

Both grids and thread blocks can have a spatial organization up to three dimensions:

```
dim3 dimBlock(Bx,By,Bz);
dim3 dimGrid(Gx,Gy,Gz);
```

but often, just a single dimension is employed. The adopted dimensions and exact sizes in each dimension play an important role because they should be closely related to a GPU specification, as well as the parallel algorithm's granularity. Depending on the hardware, the number of threads and blocks could be limited (typically, on current GPUs, up to 1024 threads per block and up to 65535 blocks per dimension).

From a programming perspective, the kernel is defined in C++ by:

```
__global__ void KernelName(Par1, Par2, ...)
{
    // thread index
    int i = blockIdx.x * blockDim.x + threadIdx.x;
    ...
}
```

and a thread index enumeration is included.

Each block is assigned to one SM, where the threads are organized into *warps* consisting of thirty-two threads. Threads in one warp run the same instruction set at the same time. When the code of the kernel needs to be executed differently in threads from the same warp (e.g., because of conditional statements based on the processed data), the performance can be degraded because different instructions are realized sequentially. This effect is called *warp divergence*. Warps are run concurrently in the SM. The GPU scheduler tries to maximize the occupancy and performance by preferring the warps that are ready for computation over the idle warps (e.g., waiting for data access or function results).

Finally, CUDA can be seen as a hierarchy of computation (the threads, thread blocks, and kernels) and the corresponding hierarchy of memory spaces (local, shared, and global) and synchronization of primitives.

A typical flow of a CUDA program (on the host side) can be realized in the following steps:

1. Allocate memory on the device (`cudaMalloc`);
2. Copy data from the host to the device (`cudaMemcpy`);
3. Run kernel(s) on the defined grid of thread blocks;
4. Copy the results from the device to the host (`cudaMemcpy`);
5. Release memory (`cudaFree`).

It is important to remember that any kernel execution needs to be preceded by an appropriate memory allocation on the device and data transfer to the device. Similarly, afterward, the results should be retrieved, and the allocated memory should be released.

The CUDA framework can be attractive, even for inexperienced programmers, because it is available as a simple extension of the C/C++ language. It offers a rather smooth start, especially for parallelization of typical tasks. NVIDIA shares a lot of highly optimized libraries that can be used without a profound understanding of the underlying mechanisms.

Because of its very competitive price-to-computation-power ratio, GPUs are recently widely applied in various large-scale data mining projects. A detailed and up-to-date survey can be found in [17], but in the case of decision trees, the main effort is concentrated on speeding up the top-down induction [19] or the ensemble learning

[20]. Regarding evolutionary computation, all major parallel and distributed models (master-slave, island, and cellular) have been realized on the GPU (see [21] for the recent overview). Concerning GPU-based evolutionary data mining, it is worth noting the works dealing with genetic programming, for example, [22] and [23].

3.6   Apache Spark

MapReduce [24] was one of the first distributed frameworks for processing large-scale data on computing clusters and was first introduced by Google. The programming model is based on a simple strategy, where data are split and processed, and the results are finally combined. For certain applications, it leads to a remarkable success because of the high scalability and fault tolerance. Apache Hadoop offers an open-source implementation of MapReduce that supports distributed shuffles. With rising usage and expectations, it becomes clear that the framework has limitations. Among its successors is Apache Spark, which seems to be the most promising approach nowadays.

Spark [25] is an open-source distributed computing engine for large-scale data processing. In contrast to Hadoop, where intermediate results are stored on disks, Spark processes data in a distributed shared memory model, preferably in the cluster nodes' RAM. Such an in-memory computing scheme is undoubtedly better suited for iterative algorithms and interactive data exploration. Spark provides high-level APIs in a few programming languages (Java, Scala, Python, and R). It encompasses the fundamental Spark Core module and specialized libraries, which offer tools for structured data processing (Spark SQL), machine learning (MLlib), graph processing (GraphX), and data streaming (Spark Streaming).

Apache Spark requires a cluster manager (it can be a native Spark manager or other, e.g., Hadoop YARN) and a distributed storage system (typically HDFS,[3] but other systems are possible).

The Spark architecture is organized around a concept of Resilient Distributed Dataset (RDD), which is an immutable, fault-tolerant collection of elements distributed over nodes that can be manipulated in parallel. Two basic operations can be carried out in Spark:

- Transformation—it creates a new RDD by transforming the original RDD, which remains unchanged. This is a lazy operation—it will not be executed until a subsequent action needs the result. Representative examples of transformations are mapping and filtering;
- Action—it processes but does not change the original data and returns the calculated values. In contrast to a transformation, this operation is not lazy. Counting is a good example of an action.

---

[3]HDFS stands for Hadoop Distributed File System.

A lazy evaluation is an interesting feature of Spark because it naturally eliminates the computations that are not needed. RDDs provide explicit support for data sharing among computations. Users can persist selected RDD in memory, and this feature is the main advantage of Spark over older MapReduce computing models, providing large speedups, especially for iterative algorithms.

Among the many possible transformations that can be applied to the RDD, let us shortly review just a few representative examples:

- `map(function)`—it returns a new RDD formed by each element of the original RDD through the `function`;
- `mapPartitions(function)`—similar to `map` but runs separately on each partition of the original RDD;
- `filter(function)`—it returns a new dataset formed by selecting those elements of the source on which `function` returns a true value;
- `union(otherRDD)`—it enables the merging of the RDDs. Returns a new RDD that contains the union of the elements in the RDDs;
- `reduceByKey(function, [numTasks])`—when called on a dataset of (`K`, `V`) pairs, it returns a dataset of (`K`, `V`) pairs where the `V` values for each key `K` are aggregated using the given reduce `function`.

Concerning the various actions available in Spark, at least two examples are worth being mentioned:

- `reduce(function)`—it aggregates the elements of the RDD using a `function` (the function takes two arguments and returns one);
- `countByValue()`—returns the counts of the unique values in the RDD as a local map of (`value`, `count`) pairs.

It should be noted that Spark provides two types of shared variables:

- Broadcast variables—reference read-only data that need to be distributed over all nodes. For example, it can be used to give every node a copy of a large input dataset in an efficient way;
- Accumulators—used to efficiently aggregate the information in program reductions (usually to implement counters and sums).

As pointed out, the predecessors of Spark were not best suited for iterative algorithms; hence, it naturally limited their applications to evolutionary computation. In [26], the authors investigate the advantages and deficiencies of applying Hadoop MapReduce for improving the scalability of parallel genetic algorithms. Because the overhead of data store, communication and latency are considerable, they conclude that only the island model with rare migrations can be efficiently realized independently of the problem's computational load.

The emergence of the Spark framework opens up new opportunities for distributed realizations of evolutionary algorithms. One of the first attempts is proposed by Deng et al. [27], and the authors develop a parallel version of differential evolution. A population is treated as an RDD, and only the fitness evaluation is distributed to workers. Differential evolution is also investigated by Teijero et al. [28], and they focus on an individual's mutation. The authors study three master-slave implementations, but the observed efficiency is not satisfactory. Hence, they switch to the island model with local-range mutations and rare migrations. In [29], a parallel genetic algorithm based on Spark for pairwise test suite generation is introduced. In this approach, a population of individuals is stored as an RDD, and a fitness function is evaluated on workers. Genetic operators are applied in sub-populations that correspond to partitions. Recently, Barba-Gonzalez et al. [30] have proposed a more general approach for the parallelization of evaluation steps in metaheuristics. Their tool combines the JMetal multi-objective optimization framework with Spark, which processes a population of individuals as an RDD.

As for applying Spark to evolutionary data mining, only a few approaches have been published so far. In the first paper [31], the authors try to scale a genetic programming-based solution for symbolic regression and propose a distributed fitness evaluation service. Ferranti et al. [32] investigate the fuzzy rule-based classifier generation from Big Data. In their approach, a fitness function of a multi-objective algorithm scans for the entire training dataset. This is obviously a computationally demanding operation, and it is distributed among cluster nodes. Another multi-objective evolutionary approach for subgroup discovery is introduced in [33]. In separate dataset partitions, the sets of fuzzy rules are extracted, and for each value of the target variable, the search is repeated. The rules obtained in all partitions are then reduced based on the proposed global measures.

References

1. Moore G (1975) Progress in digital integrated electronics. Proc Int Electron Devices Meet 21:11–13
2. Flynn M (1972) IEEE Trans Comput C–21(9):948–960
3. Owens J, Luebke D, Govindaraju N, Harris M, Krger J, Lefohn A, Purcell T (2007) A survey of general-purpose computation on graphics hardware. Comput Graph Forum 26(1):80–113
4. Jeffers J, Reinders J, Sodani A (2016) Intel xeon phi processor high performance programming: knights landing edition. Morgan Kaufmann, Burlington
5. Johnson E (1988) Comput Arch News 16(3):44–47
6. Krauter K, Buyya R, Maheswaran M (2002) Softw: Pract Exp 32(2):135–164
7. Foster I (1995) Designing and building parallel programs. Addison-Wesley, Boston
8. Grama A, Karypis G, Kumar V, Gupta A (2003) Introduction to parallel computing. Addison-Wesley, Boston
9. Eijkhout V (2015) Introduction to high performance scientific computing
10. Xu C, Lau F (1997) Load balancing in parallel computers: theory and practice. Kluwer Academic Publishers, Dordrecht
11. Amdahl G (1967) Validity of the single processor approach to achieving large scale computing capabilities. In: Proceedings of AFIPS'67. ACM, Providence, pp 483–485
12. Gropp W, Lusk E, Skjellum A (2014) Using MPI: portable parallel programming with the message-passing interface, 3rd edn. The MIT Press, Cambridge
13. Chapman B, Jost B, van der Pas R, Kuck D (2007) Using OpenMP: portable shared memory parallel programming. MIT Press, Cambridge
14. van der Pas R, Stotzer E, Terboven C (2017) Using OpenMP-The Next Step: Affinity, Accelerators, Tasking, and SIMD. MIT Press, Cambridge

15. Gaster B, Howes L, Kaeli D, Mistry P, Schaa D (2011) Heterogeneous computing with OpenCL. Morgan Kaufmann, Burlington
16. Wilt N (2013) CUDA handbook: a comprehensive guide to GPU programming. Addison-Wesley, Reading
17. Cano A (2018) WIREs Data Mining Knowl Discov 8(1):e1232
18. NVIDIA (2018) CUDA C programming guide. http://docs.nvidia.com/cuda/pdf/CUDA_C_Programming_Guide.pdf
19. Nasridinov A, Lee Y, Park Y (2014) Computing 96(5):403–413
20. Marron D, Bifet A, Morales G (2014) Random forests of very fast decision trees on GPU for mining evolving big data streams. In: Proceedings of ECAI'14. IOS Press, pp 615–620
21. Cheng J, Gen M (2019) Comput Ind Eng 128:514–525
22. Langdon W (2011) Soft Comput 15(8):1657–1669
23. Chitty D (2016) Soft Comput 20(2):661–680
24. Dean J, Ghemawat S (2008) Commun ACM 51(1):107–113
25. Zaharia M et al (2016) Commun ACM 59(11):56–65
26. Ferrucci F, Salza P, Sarro F (2018) Evol Comput. https://doi.org/10.1162/EVCO_a_00213
27. Deng C, Tan X, Dong X, Tan Y (2015) A parallel version of differential evolution based on resilient distributed dataset model. In: BIC-TA 2015 Proceedings, pp 84–93
28. Teijeiro D, Pardo X, Gonzalez P, Banga J, Doallo R (2016) Implementing parallel differential evolution on Spark. In: Proceedings of EvoApplications 2016. Lecture notes in computer science, vol 9598, pp 75–90
29. Qi R, Wang Z, Li S (2016) J Comput Sci Technol 31(2):417–427
30. Barba-Gonzalez C, Garcia-Nieto J, Nebro A, Aldana-Montes J (2017) Multi-objective big data optimization with jMetal and Spark. In: Proceedings of EMO'17. Lecture notes in computer science, vol 10173, pp 16–30
31. Funika W, Koperek P (2016) Towards a scalable distributed fitness evaluation service. In: Proceedings of PPAM'15. Lecture notes in computer science, vol 9573, pp 493–502
32. Ferranti A, Marcelloni F, Segatori A, Antonelli M, Ducange P (2017) Inf Sci 415–416:319–340
33. Pulgar-Rubior F, Rivera-Rivas A, Perez-Godoy M, Gonzalez P, Carmona C, del Jesus M (2017) Knowl-Based Syst 117:70–78

Part II
The Approach

Chapter 4
Global Induction of Univariate Trees

Meta-heuristic approaches like evolutionary algorithms are now known to be potentially very flexible and reliable optimisation and search methods [1]. However, their success in solving complex problems strongly depends on a proper identification of the aims, understanding the constraints and using domain knowledge as much as possible [2]. A decision tree induction based on training data is without doubt a very particular and difficult problem. The hierarchical and varied structure of the tree itself is a challenge, not to mention the huge diversity of the collected and analyzed datasets. This explains why many simple attempts to apply evolutionary techniques, such as genetic programming based on adopting available frameworks, are rather optimistic, but they are not continued. Moreover, most of the existing approaches are limited to only one tree type or prediction task. When researchers use more specific phenomena regarding decision trees, they encounter increasingly burdensome programming limitations of ready-made libraries. And strange workarounds are needed to realize the planned functionalities. It seems that decision tree evolution forces the development of both dedicated algorithms and implementations.

Over the last fifteen years, we have proposed a family of specialized evolutionary algorithms for the global induction of various types of decision trees. All variants were implemented in a native Global Decision Tree (GDT) system. As in typical software product line development, specific variants use the dedicated parts of the implementation but share major components. The system is highly flexible, and all algorithms are fully parameterized and can be steered by a user.

In this book, I will present a whole family in a concise and unified manner, focusing on the elements that seem to be important and/or promising, and not going into detail in the case of typical solutions. The GDT system can be applied for solving both classification and regression problems. In the case of regression problems, leaves can be associated with simple target values or more complex models, leading to a model tree. The models can be a simple linear regression or multiple linear regression. In this chapter, the main ideas of global induction of decision trees are introduced, but for clarity, only univariate trees with tests using one feature are described. In the

© Springer Nature Switzerland AG 2019
M. Kretowski, *Evolutionary Decision Trees in Large-Scale Data Mining*,
Studies in Big Data 59, https://doi.org/10.1007/978-3-030-21851-5_4

next chapter, more sophisticated oblique and mixed trees are discussed. A part of the developed genetic operators can be applied regardless of tree type, but there are also very specialized variants that should be discussed separately.

In typical data mining algorithms, much attention is usually devoted to the speed of operation and applicability to large datasets [3]. In the case of an evolutionary approach [4], it is even more evident because iterative and population based algorithms are inherently not the fastest ones. Naturally, a lot of effort is given ensuring that the GDT system is not only reliable and accurate but also fast enough. One of the frequently applied methods to focus an evolutionary search and improve its effectiveness is embedding the local search procedures into the algorithm. Such memetic extensions should be reasonably introduced because their computational complexity can be high, and the costs could exceed the potential profits. Searching for a new test in a node is a rather obvious candidate for boosting through a local optimisation. In evolutionary induction, it is applied during initialization, and certain mutation variants in a controlled manner (with relatively low probability) that are used too extensively reduce the advantages of the global search [5, 6]. Fortunately, through this mechanism, various methods from the most recognized systems can be easily combined in the GDT system to ensure high diversity.

A fitness function is at heart of each evolutionary algorithm, and it plays a crucial role in the evolutionary decision tree induction, which is clearly a difficult multi-objective problem. In the GDT system, various trade-offs in the fitness function can be a priori decided, or a Pareto-optimal front can be simultaneously searched.

A general structure follows a typical framework for an evolutionary algorithm with an unstructured population and a generational selection [7].

4.1 Representation

A typical scenario for applying a population-based meta-heuristic approach to a new domain relies on coding the possible solutions of a problem according to an imposed yet specific to a method representation. In this way, standard mechanisms (e.g., genetic operators) can be used, which undoubtedly facilitates acting. For example, in a classical genetic algorithm, fixed-size binary chromosomes are expected, whereas in genetic programming, graph-based or tree-based structures are evolved. Not always is such a way of proceeding productive, especially when coding is not straightforward and when an actual search space is not properly mapped. In the worst case, it can lead to an unnecessary magnification of a search space and a substantial downturn of an optimisation process.

A decision tree is a specific hierarchical structure, and its arrangement and size fully depend on the training dataset used for induction. The role of a node results from its position in a tree, and internal nodes operate differently than leaves. Before examining the dataset, it is usually not possible to decide which test types will be needed, and even the number of outcomes of these nominal tests can be questionable.

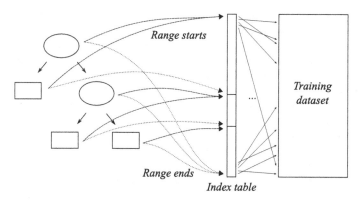

Fig. 4.1 An example of storing locations of the training data in a tree by using an index table

It is obvious that decision trees should be represented by flexible and varied-size individuals.

Another element that should be considered when deciding about decision tree representation is knowledge about the locations of all the training objects in tree nodes. Primarily, it enables fast calculations of various prediction errors that are indispensable for a fitness evaluation. Together with the other statistics maintained in nodes, it could make an informed application of genetic operators possible. Different additional data structures can be proposed for storing information about training data, and it is the dimensionality of the analyzed data that really matters.

The simplest solution requires linking each node with a separate table of training object indexes (a root node can be omitted because it corresponds to the whole dataset). Such a solution can be effective only for small- and medium-sized datasets. Large parts of tables are redundant, and tables are constantly resized and reallocated during the evolution. It should be noted here that it is a population of individuals that evolves, and this multiplies the memory requirements. Because many individuals share some subtrees as a result of a selection, a common table mechanism based on reference counting was developed [8], here trying to mitigate the memory-related problem to a certain extent. However, another less memory intensive and more stable solution can be proposed. It uses only one (full size) index table per tree, and the table does not need to be reallocated. In each node, a reference to a starting position in the table and the size of the corresponding training subset are maintained. The ranges from the child nodes are located one by one inside the range of the parent node. In Fig. 4.1, an example of this solution is shown. It should be noted that a change (e.g., because of a mutation) of a subtree causes a reorganization of a limited, continuous area in the index table. It is a very useful property because it simplifies a partial fitness recalculation to speed up the decision tree induction. A similar idea can be found in [9].

For large-scale data, providing permanent information regarding the arrangement of the training data can be too complex both in terms of memory and computation time. Hopefully, when a suitable technical infrastructure is available (e.g., GPUs), a

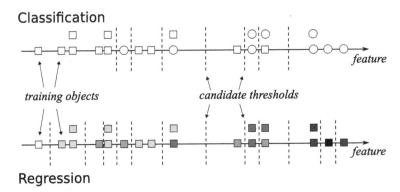

Fig. 4.2 Candidate threshold determination: the different geometric symbols represent the classes in the classification (top); the gray levels of the squares correspond to the target values (bottom). If two objects share the same feature value, the symbol of the second object is depicted above the first one

direct interaction with a learning dataset can be delegated, and it enables storing of only the necessary statistics in nodes.

Based on the above considerations, it is evident that similarities with the tree-based representations from genetic programming are rather apparent. Therefore, in the GDT system, decision trees are not specially encoded and are represented in their actual form.

In univariate trees, two test types are permitted. For nominal features, at least one feature value is associated with each branch, meaning that inner disjunction is allowed. For continuous-valued features, typical inequality tests with binary outcomes are provided, and for them, an efficient search for the proper thresholds is crucial. Fayyad and Irani [10] show that it is sufficient to consider only a finite set of potential splits for maximizing a group of general target functions in classification problems. A candidate split for a given feature can be defined as midpoint between such a successive pair of objects in a sorted sequence of feature values, such that the objects belong to different classes. This definition can be easily generalized for regression problems, but the number of splits could increase. Considering only candidate splits significantly limits the number of possible threshold values and focuses the search process. All candidate thresholds for each continuous-valued attribute are hence calculated before starting the actual evolutionary induction. In Fig. 4.2, an example of a candidate threshold determination is presented.

A user must declare a type of a predictive task for which a global induction is launched. For a classification, the leaves are associated just with a single class that is calculated according to a majority rule based on the training objects in a leaf. In the simplest case of a regression, a class is replaced with the average of a target feature. In more advanced cases, each leaf is associated with a model. It can be a simple linear model using only one independent feature or a multiple linear model that can rely on more features (up to all independent features). Models are fitted with a standard minimization of the squared residuals [11].

4.2 Initialization

Before starting an actual evolution, an initial population should be created. Individuals should be diverse, and they should cover the entire range of possible solutions because doing so enhances the efficiency of an exploration during the evolution. Usually, the initial individuals are randomly created, but in the case of decision trees, which are rather complicated hierarchical structures, starting with completely random (and often useless) trees does not make any sense. The search space of all possible decision trees, even for moderate training datasets, is immense. Especially for Big Data applications, a slighted initialization will inevitably slow down the search significantly and could lead to the failure of an evolutionary approach. On the other hand, the computational complexity of an initial tree generation algorithm should be low (or at most acceptable).

It is rather obvious that a rational method for generating a sensible initial tree uses a top-down scheme. As an elementary variant of such a tree construction, a very simple dipolar initialization can be performed [12]. A dipole is just a pair of objects, and such a pair can be easily used to find an effective test,[1] providing that the values of at least one feature are different. When an appropriate test has been found, it can be said that it splits the dipole.

In each step of a recursive procedure, one dipole is randomly chosen based on the training instances available in a node, and it is tried to generate a test. First, features with different values are identified, and one feature is randomly chosen. If a nominal feature is drawn, an equality test is directly created (with separate test outcomes associated with each possible feature value). For a continuous-valued feature, an inequality test is created; hence, an additional draw is needed for a threshold that will split the dipole. A range for the possible threshold value is derived from the feature values corresponding to two objects.

In the case of a classification problem, we are interested only in so-called mixed dipoles [13] with objects belonging to distinct classes because we want to separate the objects from different classes. So the first object can be chosen completely at random, but for the choice of the second one, the objects from the same class are excluded. For regression problems, the situation is a little bit more complicated because the constraints for a dipole choice are not so evident. We decided to prefer splitting the dipoles with a higher difference in predicted target values. The first object is simply randomly chosen, and for choosing the second one, the available objects are sorted according to the calculated differences in the target values. Then, a mechanism similar to the ranking linear selection is applied to draw the second object, with a preference to longer dipoles.

The recursive partitioning finishes when the number of training objects drops below the fixed value (default: five). It stops also when all objects in a node belong to one class (classification) or have the same or very similar target values (regression). It is rather obvious that the trees generated with such a fast but simplified method

---

[1] A test is effective for a given subset of training data if it returns at least two different outcomes (which means that it directs the objects to at least two branches).

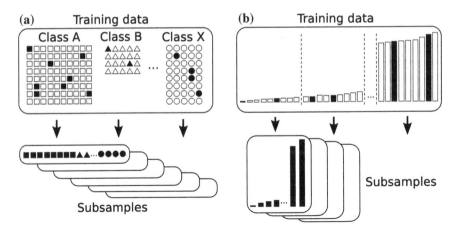

Fig. 4.3 Stratified random selection of the instances from the training data into subsamples: **a** classification and **b** regression

will be overgrown, especially for large datasets. It is why a maximum initial tree depth parameter was introduced (default: five), avoiding an extensive decision tree pruning and slowing down the first phases of an evolutionary induction. An even more restrictive idea was introduced in [14], where the initial trees are created as decision stumps with only one test.

Another mechanism that can be introduced to prevent wasting resources when the training data are vast relies on using only small subsets of the training data for initialization purposes. For each individual, a subsample of the training data is randomly chosen. The default subsample size is ten percent of the dataset but no more than 500 instances [15]. This amount of objects is fairly large enough for constructing diverse and reasonable-sized initial trees. Drawing a subsample is performed in a stratified way to ensure the representativeness of the subsamples (Fig. 4.3).

For classification data, this means preserving the class distribution from the full training dataset in each subsample. This is realized by drawing the proportional number of objects in each class. However, for regression data, we try to imitate a profile of the dependent variable. The original training dataset is sorted according to the values of the target, and instances are placed into a defined number of equal-sized folds (default: ten). Next, the same number of objects is randomly chosen from each fold to be included in a subsample.

A similar mechanism to reduce the number of objects used for initialisation can be applied to features, especially when the number of features is important, like, for example, in micro-array data [16]. For each individual, the set of features, which is taken into account when searching for tests, can be randomly limited to a fixed size (default: fifty percent of the original features) [17].

Generally, it is better to have not only diverse but also sensible starting individuals because it obviously improves the evolutionary algorithm's efficiency. In the GAIT system [18], the authors propose creating an initial population with classification trees generated by C4.5 [19] and evolving them through a number of generations to yield trees of a better quality. To initialize a genetic search with a diverse population, the subsamples of the original data are used to generate the first trees. Inspired by their idea, we decided to go further and enable initializing an evolutionary induction by trees generated with various optimality criteria [5]. In the case of classification trees, parts of the initial population can be constructed using the following well-known local goodness-of-split criteria:

- Gini index of diversity
- Information gain
- Gain ratio

Apart from homogenous tree creation (with all the tests based on the same criterion), it is possible to build an initial tree where a method for choosing a test in each node is randomly chosen. For all the aforementioned criteria, the best test in a given node is selected from the highest rated tests calculated for each feature. A procedure for an optimal test choice needs to be computationally effective. This is why for a nominal feature with many values, grouping together feature values is not considered here. In the case of a continuous-valued feature and searching for locally optimal inequality tests according to the discussed measures, an efficient method can be implemented using dedicated data structures. This requires a single pass through an ordered list of boundary thresholds with corresponding training instances. In each step, a distribution of classes on both sides of a currently processed threshold is incrementally updated.

In the case of an initialisation of regression tree induction, an analogous concept of involving goodness-of-split criteria has been introduced [6], but other measures need to be optimized:

- Least squares (LS)
- Least absolute deviation (LAD)

And similarly, both homogenous and heterogenous trees can be included in an initial population.

The initialization of model trees requires additionally fitting models in all the leaves. In the case of a simple linear model in a leaf [20], each time one feature is randomly chosen, a model is fitted using the available training instances in that leaf. It is also possible to apply a more computationally demanding method and to iterate through all the features. As a result, the best model that minimizes the optimality measure can be chosen. When multiple linear models are allowed, a set of features used for fitting a model can be restricted only to the features that were employed in the tests on a path from the root node to a leaf [21]. Such a solution considerably reduces computational complexity. It should be recalled that the number of features used in a model can be restricted by the low number of available training objects in a leaf.

The generated initial trees can be highly overgrown, which is not good for evolution because it requires wasting the computational resources to significantly reduce the tree sizes. It is much more economical to perform a starting pruning to maximize the initial fitness.

In addition, in the GDT system, the most complex model trees can be initialized with a method inspired by the M5 initialization [22]. Here, an unpruned regression tree is generated first, and then, the multiple linear models are fitted in all the nodes using the features from the tests in the corresponding subtrees. Finally, the tree is recursively pruned back, starting from the leaves, as long as the estimated error decreases.

Because each tree is generated on a small and random subsample of the training data, a method for the creation of tests can be randomly chosen from the simplest dipolar, and a few locally optimal approaches, the resulting initial population is really diverse and is a rational starting point for the evolutionary induction of decision trees.

4.3   Specialized Genetic Operators

When developing a new evolutionary algorithm for a given problem, it is natural that the design decisions related to a representation can heavily influence other decisions, especially concerning variation operators. In an evolutionary induction, because we decided to avoid encoding solutions and to retain the original form of decision trees, we have to propose specialized genetic operators. This can be a chance to reasonably focus the search in a very complex and vast search space of hierarchical and inhomogeneous structures by incorporating a deep understanding of the predictive process, for example. On the other hand, it is also a danger that the poorly defined specializations would destroy a key balance between exploration and exploitation. It also very important to remember that in the context of Big Data applications, the computational complexity of the operators need to be carefully analyzed. It is especially valid for various memetic extensions, where temptations to conduct exhaustive enumerations or calculations are ubiquitous.

For an efficient variation of the decision trees, we have proposed many variants of genetic operators that enable mutating a single individual and recombining a pair of solutions. We also have developed a consistent mechanism for controlling frequency and the context-specific application of the variants.

After applying any operator, it is necessary just to update (if all the locations are stored) or fully calculate (otherwise) the information concerning location of the training instances in the tree nodes in the affected parts of trees. This provides the essential statistics (like errors estimated on the training datatsets) for whole trees to evaluate individuals.

4.3.1   *Mutation*

In classical genetic algorithms, a mutation can be applied independently with an equal and relatively small probability to any position of a chromosome. For a fixed-sized representation, the probability can be attuned to ensure that every individual is affected on average once per generation [7]. Similar ideas can spread over many evolutionary approaches. In our first papers [12, 23], we tried to adopt such a solution for a decision tree's evolution, and every node of a tree would have the same chance of being modified. However, when applied to hierarchical, heterogeneous, and variable-sized structures, many deficiencies with this method, become evident. For example, the impact of a root node modification is huge because it results in complete tree rebuilding, whereas the modification of a node in lower parts of a tree affects only a small tree part. Moreover, tree sizes can differ significantly for a given problem and between various datasets. Typically, trees are large in the first phase of the induction, and their size decreases during the evolution. In such a scenario, if a relatively low mutation probability is established, it works well at the beginning for large trees, but for small trees in later phases, it obviously slows down the evolution because trees with just a few nodes are almost not mutated at all. An inversed scheme is not much better because large trees are mutated many times in one generation, which is usually destructive. Also, a noticeable number of mutations does not affect a tree at all. It is especially visible for leaves, where all the training instances in a node come from the same class or have very similar target values.

A new scheme for applying several specific mutation variants to various node types in a uniform and controlled manner was proposed in [24]. It aims at modifying just one node (with its subtree) in a tree with a given probability, regardless of the tree's size. Mutation variants that a priori are known to be unsuccessful for a chosen node are not even considered. Moreover, the user preferences for mutation variants can be easily introduced.

As a first step, it is decided what type of node will be mutated (a leaf or an internal node). The type is randomly chosen with an equal probability.[2] Next, a ranked list of nodes of the expected type is created, and a mechanism analogous to ranking linear selection is applied to decide which node will be processed. It should be noted that in specific conditions, the list can be empty. This most frequently occurs when every leaf contains only training objects from the same class and when any modification of a leaf cannot improve the (rationally defined) fitness. Concerning internal nodes, only a tree reduced to a single leaf poses a problem because there is no internal node to modify at all. When a mutation of the selected node type is not possible, the second type is considered.

As for leaves, accuracies estimated on the training instances in each leaf can be good properties to put these leaves in order. A rationale for deciding that leaves with higher error rates should be mutated with a higher probability is rather obvious. And homogenous leaves with training instances from one class or with the same target are excluded from this ranking. Such a strategy simply improves the odds of increasing the overall performance. The measure used to sort leaves can be absolute or average (it is divided by the number of training instances in a leaf). As a result, the rankings can be slightly different, but the general idea is preserved. For classification trees,

---

[2]In a complete binary tree, the numbers of internal nodes and leaves are almost equal.

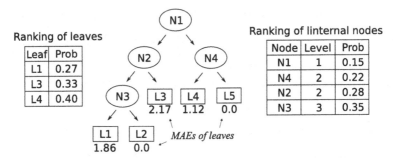

Fig. 4.4 An example of constructing the ranking lists of the leaves and internal nodes of a regression tree

the number of misclassified objects is typically used, whereas for regression trees, it is the mean absolute error (MAE).

For internal nodes, we want to primarily take into account the locations of nodes in a tree because the nodes closer to the root should be mutated less frequently. For nodes on the same level, it is still accuracy of the node that must be considered, but this time, it should be calculated for a subtree starting in the node. In Fig. 4.4, an example of building two ranking lists is presented for a regression tree.

When it is decided which node will be mutated, a list of acceptable operations is identified based on the tree type and node type.[3] Moreover, certain operations can be applied only to a specific test type. For each variant of genetic operators, before the algorithm is launched, a user can assign a relative importance value. Every time a variant to be applied is decided, all matching types are identified, and based on their assigned importance, the probabilities of the variants are calculated. Then, a simple roulette-based draw is performed, and the selected variant is realized.

For an internal node, a catalog of available operations is as follows:

1. An existing test can be modified;
2. A test can be replaced by another existing test, or a node (subtree) can be replaced by another existing node (subtree);
3. A completely new test can be generated; and
4. A node can be converted into a leaf, which is equivalent to pruning a subtree starting with this node.

Modifying a test for a nominal feature makes sense only if it is not a binary feature, and internal disjunction (grouping values into subsets) is permitted. Regrouping the values associated with branches can be realised by three atomic operations: two branches can be merged, a value can be moved from one branch to another branch, and a set of values associated with one branch can be divided, and a new branch with one value is created. In Fig. 4.5, examples of these operations are depicted.

---

[3]The same mechanism is applied for the random selection of a crossover variant.

Fig. 4.5 Mutation variants of a test for a nominal feature: **a** merging branches, **b** moving a value between branches, and **c** adding a branch

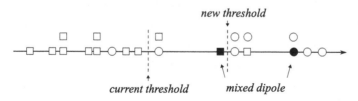

Fig. 4.6 Mutation of an inequality test: a threshold can be shifted by using a dipole as a driver

A mutation of an inequality test for continuous-valued features consists of shifting a threshold. This can be done completely at random just by drawing a new threshold from the set of candidate thresholds, or a simple dipolar algorithm [25] can be adapted. This algorithm first finds an undivided mixed (classification) or long (regression) dipole, and next, it randomly chooses a threshold from a range that will guarantee splitting this dipole. In Fig. 4.6, an example of such an operation is presented.

As for the mutation of leaves, the situation is much simpler for classification trees because a decision associated with a given leaf is fully dependent on the training instances that reached this leaf. The only meaningful mutation of a leaf consists of converting it into an internal node.

For model trees, the situation is more complicated, and it depends on the type of admitted models. If only simple linear models are considered, a mutation can exchange a feature used in a model, and then, the model can be refitted. If the memetic extensions are accepted, every feature can be considered to build the model, and the optimal one can be selected [6]. However, such an extensive search is computationally complex and should be applied with caution.

For multiple linear models, the most crucial issue is the choice of the features that are used in the model. Generally, simpler models are preferred, and in addition, the number of involved features can be limited by the small number of available training instances in a leaf. In the simplest scenario, the features can be randomly added, removed, or exchanged with a slight preference given to feature dropping. Additional variants can rest on the elimination of the least-important feature from a model or replacing a multiple linear model with a simple one (random or even optimal) [21]. In contrast to the simple linear models, for multiple models, an exhaustive search over all feature subsets is obviously computationally unfeasible.

One additional activity should be mentioned when the models in leaves are being discussed. When a subset of the training instances in a leaf is changed because of, for example, a modification in the upper part of a tree, the model in the leaf should be refitted. This means that a mutation of a test in a root node can trigger a necessity to recalculate all models in the whole tree. The same applies to decisions associated with leaves in classification trees because they are also directly dependent on the training instances.

4.3.2 Crossover

Recombination is a way of transferring or exchanging partial genetic information between trees. Potentially, it could lead to very positive results when two good individuals breed even better offspring. Unfortunately, for decision tree structures, the successful transplantation of a subtree to another tree is not an easy task. First, when a subtree is moved to a new location, its quality becomes completely unknown and can change dramatically. This is because of a context change, where the context is understood as a subset of the training instances in the starting node of the subtree. Changing a context completely often results in a total unmatch of existing tests to new training instances, and the grafted subtree has to be pruned as a consequence. It is also possible that a new subtree will match perfectly with a new context, but both theoretical considerations and experimental results show that a simple and blind application of subtrees (or only tests) exchange for decision trees too often leads to a degradation of the trees. In a framework of genetic programming based on tree-like structures, it has been observed that a context-insensitive crossover operator has a rather destructive effect on the offspring [4].

Moving any subtree to a homogenous leaf will be unproductive because the transplanted subtree will be instantly pruned. Similarly, the exchange of leaves does not make any sense for classification and regression trees because the original predictions are not transferred. These aforementioned problems with the simplest variants of decision tree crossover motivated us to develop context-aware variants that can be applied in an informed way. A prudent crossover has also been introduced recently, with a single shared object as a sufficient common context [26].

A dipolar crossover [8] has been proposed for classification trees, but this crossover concept can be easily generalized to modifying the regression trees. It uses one dipole as a kind of driver, ensuring that after moving subtree to a second tree, at least a minimal context is preserved. As a first step, an internal node in the first tree is randomly chosen, and then, the pivotal dipole can be drawn from mixed (or long) dipoles divided by a test in the node. The dipole will play a minimal context role when it comes from the first tree. This context should be preserved in the second tree. Hence, in the second tree, a node where two objects constituting the dipole are separated is searched. It requires, at the most, traversing from a root node to a leaf. Two situations are possible, and they result in different recombination endings. In the first situation, an internal node in the second tree is identified where the dipole is divided. Here, we have two subtrees matching each other in two parents, and they can be exchanged. In Fig. 4.7, an example of such an exchange is depicted.

The other situation is observed when the dipole is not divided in the second tree at all, which means that two objects of the dipole are located in the same leaf. Moving the identified subtree from the first tree to the second tree is desirable because it

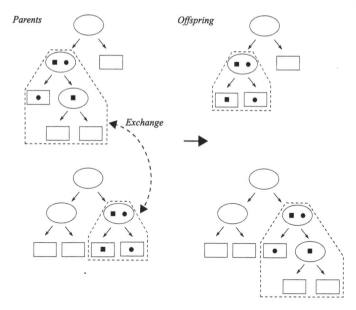

Fig. 4.7 Dipolar crossover—exchange of subtrees driven by a dipole

will result in splitting the dipole. However, a reverse move seems to be unproductive because it will just prune the target location into a leaf. Here, we try to avoid rather destructive moves, and subtree pruning is not our intention. Moreover, there is a separate variant of mutation designed for this. There are two simple solutions for this inconvenience. The order of the participating trees can be reversed. A node and a dipole are drawn from the other tree. The probability of finding a proper configuration is obviously increased, even if only a one-way move will be found once again. It is also possible to accept just one offspring.

The aforementioned problem with the second offspring and its solution becomes an inspiration for an asymmetric crossover [8], where one operation of a subtree exchange is replaced by two independent operations of copying a subtree from one tree to the other tree. It is important to note that in this type of crossover, node pairs taking part in two operations could be completely different because they are chosen apart from each other. Let us introduce the notion of a *donor* node, which starts a subtree which will be copied to another tree, and a *receiver* node, which represents a place (node) in a tree where a subtree will be inserted. As a first step, the ranked lists of possible donor and receiver places in each tree are created. Then, two pairs of donor-receiver corresponding to two copying operations are randomly selected (with the same mechanism as for a mutation). To avoid problems with consistency, two copies of the original parents are created, and they are then modified. A subtree copy from a corresponding donor node is placed in a receiver node. In Fig. 4.8, an example of two operations leading to asymmetric crossover is presented.

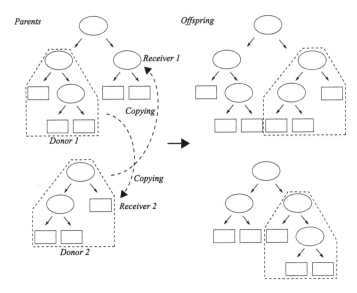

Fig. 4.8 Asymmetric crossover—two operations of copying a subtree

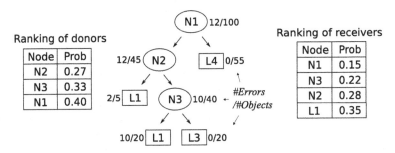

Fig. 4.9 Ranking construction for the donors and receivers used in an asymmetric crossover

A ranking construction is similar to the analogous process performed before a mutation, but other criteria have to be applied. As for donors, only internal nodes with their subtrees are considered because using a leaf as a donor results simply in pruning a receiver position. Among non-terminal nodes, subtrees with high accuracy (or low error) are preferred because they will be duplicated in the destination. In the case of receivers, we are interested in replacing worse subtrees and avoiding the destruction of the better one. This means that homologous leaves are excluded from the ranking. Moreover, leaves with very few objects are also omitted because inserting them in such positions, even in very expanded trees, will finally lead to extensive pruning, and nothing will change. In Fig. 4.9, an example of a ranking construction for a classification tree is depicted.

4.4 Fitness Function

An elaboration of very sophisticated genetic operators is not enough to develop an efficient evolutionary algorithm for a specific problem. A key factor for the successful application of an evolutionary approach is a proper definition of a fitness function. The impact of an incorrect evaluation of individual fitness cannot be compensated by other components, as it can be observed to a certain extent between recombination and mutation. For a prediction, which is the most typical data mining task, describing the expectations for an ideal system is easy. A common user is interested in a system that accurately (or even perfectly) proposes decisions for new, unseen earlier instances. And the effectiveness of any predictive system can be precisely measured after using it in an operative scenario. However, such a post-factum measure cannot be applied to drive an evolutionary search because during learning, only the training dataset is available. It is well-known that an estimation of a classifier quality on the training data typically leads to an overoptimistic result and is strongly related to the overfitting problem [27]. When considering decision trees, an extreme situation can be recalled here, one where all the leaves in a tree are associated with single training objects. In both classification and regression cases, an error estimated on the training dataset will be equal to zero, but the generalization ability of such an overgrown tree is low. This results in poor performance on new instances, which here are different from the training ones. As can be expected, a reduction in tree size can decrease the performance on the training data but also could improve the generalization and the actual performance. It becomes clear that not only the estimated correctness of predictions, but also the complexity of predictive structures, should be considered in the fitness evaluation. Moreover, it should be noticed that in many real data mining applications, the interpretability of predictions is a very important issue. Without a doubt, shorter and simpler decision rules are easier to understand and may be preferred by domain experts, especially these rules do not practically affect the overall performance. Finding a trade-off between the generalization and interpretability is not a trivial task, and we will return to this problem repeatedly, especially when a multi-objective version of the fitness is discussed.

In classical evolutionary approaches, a fitness evaluation simply assigns a single value to an individual, so all the potential participatory objectives need to be combined. In the simplest method, a weighted sum of two (or more) objectives is constructed. It is called a weight formula [28]. Another approach that considers the conflicting objectives without changing the rest of the evolutionary algorithm is known as a lexicographic analysis. It analyses the objective values one by one based on the established priorities. It also requires defining certain dummy thresholds but avoids adding up non-commensurable measures like a tree error and tree size. The last approach is based on a Pareto-dominance, and it firmly changes the way of finding a solution by applying a multi-objective evolutionary optimization. Instead of searching for a single best solution in one run of an algorithm, a group of Pareto-optimal solutions is found, and then, the final choice can be found according to the preferences of the user. All these three approaches have been investigated in the framework of the global induction of decision trees and will be discussed in the subsequent sections.

4.4.1 Unified Objective

The general form of a fitness function that is minimized is as follows:

$$Fitness(T) = Err(T) + \alpha * (Compl(T) - 1.0), \qquad (4.1)$$

where $Err(T)$ is a performance measure of a tree T estimated on the training dataset, $Compl(T)$ is a measure of a tree's complexity, which should encompass both tree size and the complexity of the nodes. A value of the α parameter should be provided by a user, and this scaling parameter finds a trade-off between the two components of the fitness. In a certain sense, $\alpha * Compl(T)$ can be treated as a penalty term, which works against an over-fitting to the training data. For classification trees, a misclassification error calculated as a fraction of incorrectly predicted instances to all training objects can be used as $Err(T)$. In the case of a perfect reclassification, where all training instances are correctly predicted, $Err(T)$ is equal to zero. In the worst case, when all instances are misclassified, $Err(T)$ equals one. For regression problems, a prediction quality could be measured by a mean absolute error (MAE) or a root mean squared error (RMSE). Both measures have a minimum equal to zero, but the maximum values are not bounded, which may make it more difficult to set a proper value of α. Fortunately, maximum errors can be estimated for sensible evolving trees, and a given training dataset and relative measures can replace these original errors. Maximal errors are observed when trees are reduced to single leaves. Every effective split creates smaller training subsets in leaves, and more accurate (not worse) models can be fitted. A relative error for a given training dataset is obtained by dividing a classical error (e.g., $MAE(T)$) by the maximal value estimated on this dataset (MAE_{Max}):

$$Err_{MAE}(T) = \frac{MAE(T)}{MAE_{Max}}. \qquad (4.2)$$

The $Compl(T)$ component should be defined according to tree type. For classification trees and regression trees without models, it can be just a tree size expressed as a number of nodes ($S(T)$). The simplest tree is composed of only one leaf, a root, so the complexity equals one. For such a situation, an influence of the complexity is canceled regardless of the α value by subtracting one in Eq. 4.1. For model trees, a complexity term should account for the model complexities in the leaves. It seems that the most natural way of expressing a complexity of a regression model is to count the used features (regressors). Hence, for a leaf with a mean value, such a measure will be equal to zero, for a leaf with a simple linear model, it will be one, and for a multiple linear model with five regressors, it will be five. The overall complexity of the models in a tree can be calculated as the sum of all model complexity in it and denoted as $MS(T)$. The $Compl(T)$ for a model tree can be defined as the sum of the tree size and the overall model complexity:

$$Compl_{MT}(T) = S(T) + MS(T). \tag{4.3}$$

Here, treating (and counting) nodes and features in the same way can be debated, especially when the number of available features is high. A scaling factor can be introduced as a kind of remedy, but it increases the number of parameters, which should be tuned for a given dataset.

In statistical model selection, where the fitting is carried out by the maximization of a log-likelihood, the Akaike information criterion (AIC) [29] or the Bayesian information criterion (BIC) [30] can be applied. The AIC is typically defined as follows:

$$AIC = 2 * Par - 2 * \ln(\hat{L}), \tag{4.4}$$

where $\hat{L}$ is the maximized value of the likelihood function of the assessed models, and Par is the number of parameters estimated by the models. A definition of the BIC is only slightly different:

$$BIC = \ln(N) * Par - 2 * \ln(\hat{L}). \tag{4.5}$$

The first terms in both equations can be seen as a kind of a complexity penalty for overparametrization.

According to [31], there is no clear choice between the AIC and BIC for model selection purposes. The AIC tends to choose models that are too complex when the data size grows, whereas for finite samples, the BIC often chooses models that are too simple because of its heavy penalty for complexity.

In the TARGET system [32], the authors propose applying the BIC-based fitness function for evolving simple binary regression trees. Assuming constant variance within the terminal nodes, the maximum of the likelihood function can be expressed as follows:

$$\ln(\hat{L}) = -\frac{N}{2}[\ln 2\pi + \ln \frac{RSS(T)}{N} + 1], \tag{4.6}$$

where $RSS(T)$ is the residual sum of the squares of a tree T. The effective number of parameters $Par(T)$ of a tree T can be (conservatively) set to account for estimating the constant error variance term and a mean parameter within each of the terminal nodes:

$$Par(T) = NL(T) + 1, \tag{4.7}$$

where $NL(T)$ is the number of leaves in a tree T. In their next paper [33], the parameter number is further increased to account for test constructions in non-terminal nodes.

In our works, we also decide to investigate information criteria-based fitness functions for a regression and model tree induction. In [20], the AIC-based fitness function is introduced and then in [17], we start to apply the BIC-based approach for a model tree induction. It should be noted that for model trees, both a tree size and a sum of model complexities should be considered for defining the effective number of parameters $Par(T)$. This means that $Par(T)$ can be equal to $Compl_{MT}(T)$.

However, the fitness function construction in the weight formula and information criteria approaches are somehow similar because in all these methods, two terms are involved: one dealing with error/accuracy and one with tree complexity/parametrization.

4.4.2 Lexicographic Analysis

In most multi-objective problems, it is rather obvious that certain objectives are treated as more important than others and that the objectives can be ordered according to their importance. The major (or preferred) objective is analyzed first to make a decision about the advantage of one solution over another. Usually, the small differences can be treated as not meaningful, and the next objective is studied. This mechanism is just adopted in a lexicographic analysis. More precisely, the objectives are ordered, and for each objective, a tolerance threshold value is introduced. The threshold values can be different and can account for a specificity of an objective. For each objective, it is first analyzed if the objective value difference is greater than the corresponding threshold, and only if the difference is high enough a comparison of the objective values conclusive. Otherwise, the next objective is similarly analyzed. It is also possible that the comparisons of all objectives will not give a decision, and then, all the thresholds are divided into half, and the described procedures are repeated starting from the first objective. If the initial thresholds are not set carefully, additional repetitions may be necessary. From an algorithmic point of view, it is advisable to exclude the objectives with the same values each time prior to lunching the procedure. These comparisons in a pseudo-code look like the following:

```
Tree Compare(Tree T1, Tree T2) {
  // Ob - array of objective functions
  // Th - array of threshold values
  // NO - number of objectives
  // MAX - maximum number of repetitions

  j=0;
  while (j<MAX) {
    i=1;
    while (i<=NO)
      if (|Ob[i](T1)-Ob[i](T2)|>Th[i])
        ( Ob[i](T1)<Ob[i](T2) ? return T1 : return T2 )
      else {
```

```
        Th[i]=Th[i]/2.0;
        i++;
      }
    j++;
  }
  return NULL;
}
```

For a classification tree induction, two objectives can be analyzed: reclassification error as the first and the number of nodes as the second. In the case of model trees, the first objective should be replaced by a root squared error (RSE), and a third objective can be introduced, here dealing with the complexity of the models in the leaves. This can be measured by the overall sum of the used features in all models.

The procedure for comparing solutions is applied to create a ranking of individuals according to their fitness. And the ranking is then used in a selection to generate a new population.

In the context of the evolutionary induction of model trees, using the lexicographic analysis was first proposed by Barros et al. [34]. Based on their work, we decide to investigate such an approach in [21] and introduce such an alternative into the GDT system.

4.4.3 Pareto-Optimality

Both previously described ways of calculating the fitness function are based on some arbitrary assumptions regarding identified objectives. They can lead to solutions that do not meet analyst expectations. It can be especially adverse when the attacked problem is novel. In such a situation, an opportunity to study the various trade-offs between major objectives could be very beneficial. Fortunately, evolutionary computations could be excellent tools for this type of multi-objective optimization, as detailed in Chap. 1. And particularly in data mining, multi-objective evolutionary algorithms are being applied more and more often [35].

One of the first applications of the multi-objective evolutionary approach for searching of Pareto optimal decision trees is presented by Zhao [36]. He develops an interactive system for a cost-sensitive binary classification based on strongly typed genetic programming with simplistic operators. A rank of relative non-dominance is used as a fitness function. A user can specify partial preferences on two conflicting objectives (e.g., sensitivity versus specificity). Another Pareto-optimality-based approach for classification tree induction is presented in [37]. It extends the greedy OC1 induction of oblique trees by changing the original hill climbing and perturbation algorithms into evolutionary process. The bi-objective method minimizes the tree size and maximizes the classification accuracy. The hyper-volume metric [38] is used for evaluating the quality of a non-dominated set of solutions. In the last solution, the authors [39] study two-objective tree induction based on genetic programming. A misclassification rate estimated with five-fold cross-validation and a sum of the costs of all the features contained in a tree are used as fitness functions in a selection analogous to the non-dominating sorting procedure of NSGA-II [40]. A fixed maximum tree size safeguard is also used to prevent excessive tree growth. An interesting crossover variant based on proposed variable importance measure is introduced, and it applies a kind of a ranking selection for choosing a crossover point. Features from tests in nodes that are frequent in trees constituting a current population and features in nodes that are closer to roots can be selected with a higher probability. Finally, it is worth noting that sophisticated, visualisation tools are currently developed for an interactive and sensitivity-aware selection of Pareto-optimal decision trees [41].

In the GDT system we decide to follow the general ideas of the NSGA-II workflow to enable the generation of a set of Pareto-optimal decision trees. These extensions affect mainly the fitness function calculation and the selection mechanism. Particular detailed components are based on more recent works, and certain special solutions are proposed to better fit decision tree induction. For example, for an efficient non-dominated sorting, the ENS strategy, which was presented in Chap. 1, is applied. Similarly, the updated crowding distance procedure is used, which eliminates a problem where there are individuals sharing the same objective values. Additionally, we decide to maintain a separated, elitist archive of all the individuals in the first Pareto front because we want to reduce the probability of the overlooking possibly interesting decision trees. This results in having to reorganize the selection; the proposed method is described in the corresponding section.

It may seem that with all the non-dominated solutions archived, the impact of the crowding distance calculation would be reduced. However, the crowding mechanism plays a very important role in preserving the population diversity and thus faster constitution of the Pareto front. From an analyst perspective, a proper definition of the crowding distance could have a strong impact on the visualization of the final front and a decisive influence on the analyst's choice. In [42], we test different crowding distance rankings. For example, by incorporating the weights assigned to objectives in the ranking calculation, the analyst can better express its preferences and focus only on the most promising solutions.

In the GDT system, up to three objectives can be considered because the visualization of the Pareto front in more dimensions is rather cumbersome. For classification trees, they can be, for example, the accuracy, the tree size, and the average test complexity (if multivariate tests are allowed). For model trees, the average model size can be the desirable objective as well.

4.5 Post-Evolution Processing

The candidate thresholds precalculated for each continuous-valued feature allow focusing and accelerating the evolutionary search. In lower parts of the resulting tree, the number of training instances in the nodes is reduced, and the thresholds derived from the whole training dataset could be located in unoptimal positions. Hence, for all final inequality tests in the internal nodes, the thresholds can be shifted to a half-distance between the locally available feature values [15]. Such an operation does not change the fitness corresponding to the final tree but improves the generalization and stability.

Smoothing for model trees could be enabled (with smoothing constant $k = 10$) by the user. However, it should be noted that smoothed model trees are different from classical model trees. Prediction relies not only on a model from a leaf, but also integrates additional models from internal nodes on a path from a root to the leaf. The interpretation and understanding of such an operation are clearly more difficult.

4.6   Selection and Termination

In almost all the variants of a global induction, a classical generational selection is performed based on the linear ranking selection, which ensures a constant balance between exploration and exploitation. Moreover, in each iteration, the best individual found so far (with the highest value of the fitness function) is included into a newly created population, which means that an elitist strategy is adopted.

Only for the Pareto-based multi-objective induction [42] does another approach need to be introduced. First of all, an archive of all non-dominated solutions is maintained [43]. In NSGA-II, when a population is small, many solutions that may be interesting from an analyst's point of view can be overlooked. Solutions from the Pareto front are stored in the archive, which is updated each time a new solution from the current population dominates one in the archive. This increases the computational complexity slightly, but it seems that for decision trees, where Pareto fronts are not very large, it is profitable. Maintaining the archive with non-dominated solutions enables the application of the strategy proposed in [44], where instead of the single elitist individual, a pool of NE (default: half of the population size) elitist solutions is preserved in a new population. The rest of the new population ($P - NE$) is filled with individuals chosen from a set of parents and newly created offspring. For both separate population parts, a binary selection is applied, and individuals are scored with the crowded-comparison operator like in the original NSGA-II algorithm, but for comparing elitist solutions, only the crowding distances matter.

An evolutionary induction finishes when over a fixed number of generations a fitness of the best individual does not improve (default: 1000 iterations) because this is typically interpreted as a convergence sign of evolutionary algorithms. For very complicated and/or unusual datasets, it is possible that a convergence will be very slow, so to limit the computation time and prevent risk of unacceptably long induction, an additional condition is verified, and it specifies the maximum number of generations (default: 10,000 iterations).

4.7   Experimental Results

Much in the way of experimental validation of the global induction of decision trees has been performed and published[4] in our papers. Obviously, it does not make sense to summarize them all, but it is worth recalling selected elements to understand well the advantages and benefits of this approach, as well as its deficiencies.

4.7.1   Analytically Defined Problems

Analytically defined problems are often treated as a good starting point for introducing new methods or presenting a particular situation when the desired behavior can be demonstrated. The main advantage of using such datasets is that the expected solution is known, and the correctness of the obtained trees can be easily verified. Furthermore, the dataset size can be easily scaled, and the problem difficulty can be controlled.

Two problems along with their related datasets, will be discussed here, and the decision trees generated by the GDT system will be visualised in details. Both datasets are only two-dimensional, and the internal structures of the problems can be easily depicted. These two problems will be used in the following chapters (e.g., for the investigation of parallel implementations).

The first one (*Chess3x3*) is a classification problem with two classes, and it is a kind of generalization of the *chessboard* [12] problem presented in Fig. 2.10a. This variant even looks more difficult, but it is better suited for the top-down inducers[5] because it is less uniform than the $2x2$ version. It was chosen because the optimal tree has a moderate size (eight leaves). In Fig. 4.10, the problem is visualized with explicitly drawn decision borders, and an example of a generated decision tree using the GDT system is presented.

Here, the generated tree is (almost) perfect. The subtle inaccuracies in the threshold values are justified by a relative sparse sampling in the training dataset (only 1000 instances). The accuracy estimated on the test dataset is equal to 99.8%.

The second problem (*Armchair* [17]) is devoted to the investigation of regression tasks. It is also slightly more complicated (it requires at least four leaves with four models) than the problem in Fig. 2.10b, which is difficult for top-down inducers. In Fig. 4.11, the problem is visualized in 3D, and an example of the generated decision tree by the GDT system is also presented.

The obtained tree has an optimal size, and both tests in the internal nodes and models in leaves are very precisely generated. The estimated prediction accuracy is unquestionable ($RMSE < 0.01$). The tree constructed by the GDT system can

---

[4]Regarding the global induction of classification trees, the most detailed experimental validation can be found in [8].

[5]C4.5 returns a proper decision tree with eight leaves, and the estimated accuracy is more than 99.5%.

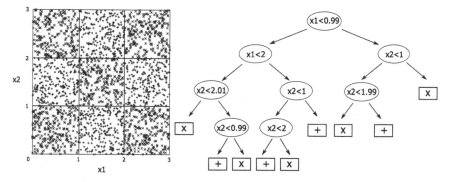

Fig. 4.10 The *Chess3x3* dataset (on left) visualization with decision borders and the corresponding example of a univariate classification tree generated using the GDT system (on right)

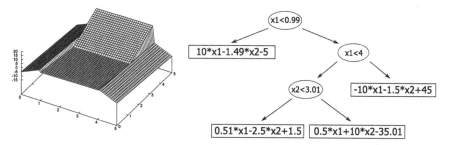

Fig. 4.11 The *Armchair* dataset visualization (on left) and the corresponding example of a univariate model tree generated by the GDT system (on right)

be confronted with two popular top-down inducers available in the Weka system [45]: REP[6] as a representative of the regression tree and M5 as the-state-of-the-art model tree inducer [22]. This dataset seems to be relatively easy, but the top-down approaches have difficulties in finding a proper first test, and as a result, the trees are overgrown. M5 started with the first split at $x1 = 3.73$ and created a tree with eighteen models in the leaves ($RMSE = 2.33$). Similarly, REP proposed the first data partition at $x1 = 4.42$, and the tree size come in at 87, but the bigger size can be justified by the simplified representation of the predictions in the leaves ($RMSE = 1.48$).

4.7.2 Unified-Objective Fitness for Regression Problems

In a typical scenario, a single-objective evolutionary algorithm is applied. Here, the results obtained for the most complex model tree induction are given. The BIC-based formula for the fitness function calculation is used. Besides M5 and REP that are used

---

[6]REP Tree builds a regression tree using variance and post-prunes it with reduced-error pruning (with back-fitting).

Table 4.1 The biggest regression dataset characteristics

Dataset	Instances	Features
Layout	66615	31
Colorhistogram	68040	31
Colormoments	68040	8
Cooctexture	68040	15
Elnino	178080	9

Table 4.2 The performance of four inducers on the biggest regression datasets

Dataset	GDT	M5	REP	Boosting
Layout	0.019	14.15	63.56	15.83
Colorhistogram	0.009	20.63	39.67	18.88
Colormoments	15.24	16.03	17.95	15.41
Cooctexture	0.964	1.225	5.55	1.00
Elnino	20.20	20.43	20.75	16.84

for single tree systems, the performance of GDT is also compared with the boosted M5 inducer (stochastic gradient boosting [46]), here being a representative of more complex (multi-tree) ensembles. The results obtained for the biggest (with more than 50000 of instances) five real-life datasets, which are analyzed in [21], are presented. The datasets were originally provided by Torgo [47] and the UCI repository [48], and their characteristics are summarized in Table 4.1.

In Table 4.2, the averages of ten runs are presented, and the performance was estimated on independent test datasets. Relative mean absolute error (RMAE) based on root mean squared error (RMSE) is used as a quality measure.[7] It should facilitate the inducer comparison because the calculated RMSE can be quite different for various datasets. The default values of the parameters for all considered systems are applied.

Here, the GDT performed better than the top-down inducers for all these datasets, and for the two first datasets, the differences are remarkable. Interestingly, only for the *Elnino* dataset does the ensemble approach show a better result. The tree complexity is also investigated for single tree inducers, and in Table 4.3, the number of leaves and the average size of the models are presented.

As can be expected, the globally induced trees are smaller, and the differences are noticeable. For the first two datasets, the GDT generated very simple trees with one test and two models in the leaves, whereas the top-down approach induced trees consisting of hundreds of leaves. Only for the *Cooctexture* dataset does the

---

[7]RMAE is equal to zero when the prediction error is zero. When the prediction returns the global RMSE mean for the dataset, the RMAE value equals 100 (%).

Table 4.3 The tree complexity (number of leaves) of three inducers on the biggest regression datasets. The average model size is provided in brackets when applicable

Dataset	GDT	M5	REP
Layout	2 (8.1)	244 (32)	118
Colorhistogram	2 (32)	395 (30)	51
Colormoments	15 (5.7)	101 (9)	630
Cooctexture	36 (11.9)	5 (16)	432
Elnino	118 (5)	877 (9.7)	2396

Table 4.4 Characteristics of the selected datasets for a demonstration of Pareto-optimal tree induction

Dataset	Instances	Features
Ailerons	13750	40
Delta Ailerons	7129	5
Delta Elevators	9517	6
Kinematics	8192	8
Pole	15000	48

M5 algorithm propose a simpler tree, but the resulting prediction accuracy is lower. Concerning the model sizes, the observed differences are not so important.

It should be stated that for the largest analyzed datasets, the evolutionary induction takes hours on a standard personal computer, whereas for the top-down counterparts, minutes are enough. It is a price that must be paid for possible more accurate and easier to interpret predictors. It also reveals why the boosting of the global approach is expected.

Much more detailed experimental validation (more datasets and systems, statistical analysis, and also a comparison with other evolutionary inducers) of the global induction of model trees can be found in [21].

4.7.3 Pareto-Optimal Regression and Model Trees

To demonstrate an alternative flow of decision tree induction based on the multi-objective approach, the results obtained for regression and model tree induction are provided. Five publicly available, real-life datasets from Torgo repository [47] are analysed. Datasets with more than 5000 instances are selected, and if the testing set is not explicitly defined, the available instances are randomly divided into the training set (66.7%) and the testing set (33.3%). The dataset characteristics are presented in Table 4.4.

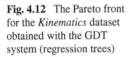

Fig. 4.12 The Pareto front
for the *Kinematics* dataset
obtained with the GDT
system (regression trees)

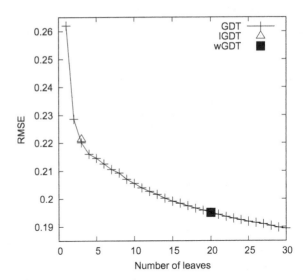

For regression trees, two objectives are considered: the prediction error measured by RMSE and the tree size expressed as the number of leaves. In the case of model trees, the third objective concerns the complexity of the models in the leaves. This is quantified by a sum of the features used in all the models. In Fig. 4.12, the Pareto front for the *Kinematics* dataset is presented. Additionally, two solutions obtained with the weighted and lexicographic fitness are presented for reference. Interestingly, these two decision trees are very distinct (three nodes versus twenty nodes). The Pareto front nicely shows a smooth transition between these two solutions, and it clearly reveals the advantage of the multi-objective approach.

In Table 4.5, the results obtained with various variants of global induction are confronted with the REP Tree (regression tree) algorithm. From many available Pareto solutions, two examples are chosen. The first one focuses on prediction accuracy (denoted as GDT[Acc]), and in the second one, tree simplicity is preferred (GDT[Sim]). Moreover, the results of the weighted and lexicographic fitness variant are included (denoted wGDT and lGDT correspondingly).

Here, the GDT[Acc] solutions indeed obtain the lowest errors among the inducers for all the datasets. The tree sizes (of all GDT variants) are clearly lower compared with the trees induced by the REP Tree. If an analyst is mostly interested in understanding predictions, the GDT[Sim] solutions offer really simplified trees with reduced accuracy. Concerning the weighted and lexicographic fitness variants, the wGDT version gives rather balanced results, whereas the lGDT version obviously prefers small trees.

In Fig. 4.13, the Pareto front for the *Kinematics* dataset is once again presented, but this time, model trees are induced. Three objectives are visualised.

It is worth noting, that the lGDT variant returned a tree, which does not even belong to the Pareto front produced by the GDT. Such a situation is much more probable

Table 4.5 Result comparison for regression tree induction. For each inducer and dataset, both the RMSE (first value) and number of leaves (the second value) are given. To increase the readability, the RMSE results are multiplied by $10^3 - 10^6$ correspondingly

Dataset	GDT[Acc]	GDT[Sim]	wGDT	lGDT	REP Tree
Ailerons	195/73	273/4	207/32	243/11	203/93
Delta Ailerons	173/66	188/8	180/13	195/4	175/251
Delta Elevat.	149/45	160/6	155/14	166/4.9	150/229
Kinematics	179/61	216/4	195/20	223/2.8	194/819
Pole	798/57	1096/15	816/58	1210/19	826/223

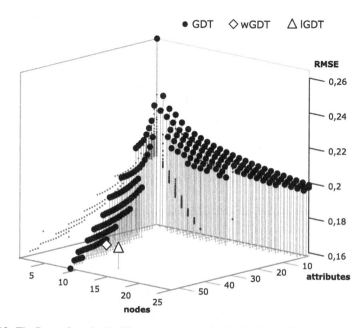

Fig. 4.13 The Pareto front for the *Kinematics* dataset obtained with the GDT system (model trees)

with three objectives considered, and it once again confirms the simplifications and restrictions of the lexicographic analysis-based approach.

The detailed results obtained for model trees are presented in Table 4.6. They are confronted with the model tree M5 system. Like, in the preceding experiment, two focussed GDT solutions are also chosen and denoted, respectively.

The GDT[Acc] variant is very competitive regarding the first criterion (RSME), but it needs quite a lot of features. Only for the *Pole* dataset does the lexicographic variant give a smaller error. Interestingly, the GDT[Sim] solutions can be very compact (mostly just one leaf with a few features in the models), and their predictions are not bad (only for *Pole* is the error visibly worse). The differences among the inducers with very similar quality of prediction can be noticeable in the number of nodes and

Table 4.6 Result comparison for model tree induction. For each inducer and dataset, three objectives are presented: RMSE (first value), number of leaves (the second value), and number of features (third value). To increase the readability, the RMSE results are multiplied by $10^3 - 10^6$ correspondingly

Dataset	GDT[Acc]	GDT[Sim]	wGDT	lGDT	M5
Ailerons	164/4/27	175/1/6	165/3.0/17	177/3.2/11	169/9/149
Delta Ailerons	167/9/29	175/1/3	173/4.6/9.5	181/4.9/12	170/17/85
Delta Elevat.	138/6/24	142/1/5	148/1.0/4	142/2.0/5.2	148/2/10
Kinematics	148/17/116	184/2/13	163/7.1/34	154/9.7/4.5	162/106/848
Pole	735/20/157	1690/3/6	807/37/52	731/78/82	749/193/1568

features. For example, for the *Pole* dataset, GDT[Acc] needs twenty nodes with 157 features, whereas M5 uses 193 nodes and more than 1500 features in the models.

A more detailed experimental analysis of the multi-objective global induction can be found in [42], where also examples of fitting a Pareto front based on various analytical preferences are presented.

References

1. Gendreau M, Potvin J (2010) Handbook of metaheuristics. Springer, Berlin
2. Michalewicz Z, Fogel D (2004) How to solve it: modern heuristics, 2nd edn. Springer, Berlin
3. Fayyad U, Piatetsky-Shapiro G, Smyth P, Uthurusamy R (1996) Advances in knowledge discovery and data mining. AAAI Press, Menlo Park
4. Freitas A (2002) Data mining and knowledge discovery with evolutionary algorithms. Springer, Berlin
5. Kretowski M (2008) A memetic algorithm for global induction of decision trees. In: Proceedings of SOFSEM'08. Lecture notes in computer science, vol 4910, pp 531–540
6. Czajkowski M, Kretowski M (2012) Does memetic approach improve global induction of regression and model trees? In: Proceedings of ICAISC'12. Lecture notes in artificial intelligence, vol 7269, pp 174–181
7. Michalewicz Z (1996) Genetic algorithms + data structures = evolution programs, 3rd edn. Springer, Berlin
8. Kretowski M (2008) Obliczenia ewolucyjne w eksploracji danych. Globalna indukcja drzew decyzyjnych, Wydawnictwo Politechniki Bialostockiej
9. Kalles D, Papagelis A (2010) Soft Comput 14(9):973–993
10. Fayyad U, Irani K (1993) Multi-interval discretization of continuous-valued attributes for classification learning. In: Proceedings of IJCAI'93. Morgan Kaufmann, pp 1022–1027
11. Press W, Teukolsky S, Vetterling W, Flannery B (2007) Numerical recipes: the art of scientific computing, 3rd edn. Cambridge University Press, Cambridge
12. Kretowski M, Grzes M (2005) Global learning of decision trees by an evolutionary algorithm. In: Saeed K, Pejas J (eds) Information processing and security systems. Springer, pp 401–410
13. Bobrowski L (1996) Piecewise-linear classifiers, formal neurons and separability of the learning sets. In: Proceedings of 13 ICPR. IEEE Computer Society Press, pp 224–228
14. Papagelis A, Kalles D (2001) Breeding decision trees using evolutionary techniques. In: Proceedings of ICML'01. Morgan Kaufmann, pp 393–400
15. Kretowski M, Grzes M (2007) Int J Data Wareh Min 3(4):68–82

16. Grzes M, Kretowski M (2007) Biocybern Biomed Eng 27(3):29–42
17. Czajkowski M, Kretowski M (2011) An evolutionary algorithm for global induction of regression trees with multivariate linear models. In: Proceedings of ISMIS'11. Lecture notes in artificial intelligence, vol 6804, pp 230–239
18. Fu Z, Golden B, Lele S, Raghavan S, Wasil E (2003) INFORMS J Comput 15(1):3–22
19. Quinlan J (1993) C4.5: programs for machine learning. Morgan Kaufmann, San Francisco
20. Czajkowski M, Kretowski M (2010) Globally induced model trees: an evolutionary approach. In: Proceedings of PPSN XI. Lecture notes in computer science, vol 6238, pp 324–333
21. Czajkowski M, Kretowski M (2014) Inf Sci 288:153–173
22. Quinlan J (1992) Learning with continuous classes. In: Proceedings AI'92, pp 343–348
23. Kretowski M, Grzes M (2005) Global induction of oblique decision trees: an evolutionary approach. In: Proceedings of IIPWM05. Springer, pp 309–318
24. Kretowski M, Grzes M (2006) Evolutionary learning of linear trees with embedded feature selection. In: Proceedings of ICAISC'06. Lecture notes in artificial intelligence, vol 4029, pp 400–409
25. Kretowski M (2004) An evolutionary algorithm for oblique decision tree induction. In: Proceedings of ICAISC'04. Lecture notes in artificial intelligence, vol 3070, pp 432–437
26. Sprogar M (2015) Genet Prog Evolvable Mach 16:499
27. Duda O, Heart P, Stork D (2001) Pattern classification, 2nd edn. Wiley, New York
28. Freitas A (2004) ACM SIGKDD Explor Newsl 6(2):77–86
29. Akaike H (1974) IEEE Trans Autom Control 19:716–723
30. Schwarz G (1978) Ann Stat 6:461–464
31. Hastie T, Tibshirani R, Friedman J (2009) The elements of statistical learning: data mining, inference and prediction, 2nd edn. Springer, Berlin
32. Fan G, Gray JB (2005) J Comput Graph Stat 14(1):206–218
33. Gray J, Fan G (2008) Comput Stat Data Anal 52(3):1362–1372
34. Barros R, Ruiz D, Basgalupp M (2011) Inf Sci 181:954–971
35. Mukhopadhyay A, Maulik U, Bandyopadhyay S, Coello C (2014) IEEE Trans Evol Comput 18(1):4–19
36. Zhao H (2007) Decis Support Syst 43(3):809–826
37. Pangilinan J, Janssens G (2011) J Glob Optim 51:301–311
38. Zitzler E, Thiele L (2000) Evol Comput 8:125–148
39. Casjens S, Schwender H, Bruning T, Ickstadt K (2015) J Heuristics 21:1–24
40. Deb K, Pratap A, Agarwal S, Meyarivan T (2002) IEEE Trans Evol Comput 6(2):182–197
41. Muhlbacher T, Linhardt L, Moller T, Piringer H (2018) IEEE Trans Vis Comput Graphs 24(1):174–183
42. Czajkowski M, Kretowski M (2019) Soft Comput 23(5):1423–1437
43. Zitzler E, Thiele L (1999) IEEE Trans Evol Comput 3(4):257–271
44. Ishibuchi H, Murata T (1998) IEEE Trans SMC, Part C 28(3):392–403
45. Frank E, Hall M, Witten I (2016) The WEKA workbench. Online appendix for "Data mining: practical machine learning tools and techniques", 4th edn. Morgan Kaufmann, Burlington
46. Friedman J (2002) Comput Stat Data Anal 38(4):367–378
47. Torgo L (2018) Regression datasets repository. http://www.dcc.fc.up.pt/~ltorgo/Regressio/DataSets.html
48. Dua D, Karra Taniskidou E (2017) UCI machine learning repository. University of California, School of Information and Computer Science, Irvine, CA. http://archive.ics.uci.edu/ml

Chapter 5
Oblique and Mixed Decision Trees

The most popular decision tree systems are based on greedy, top-down induction, and they mainly use univariate tests. The trees generated by such systems are frequently overgrown and unstable [1]. Globally induced decision trees are smaller, as shown in the previous chapter, but still, for certain problems, especially when decision borders are not axis-parallel, room for improvement exists. However, this requires making a decision tree representation more flexible by introducing oblique tests in non-terminal nodes.

Building a test based on a hyperplane is a much more computationally complex task. First, there is no way to precalculate potential solutions like with candidate thresholds for inequality tests. A splitting hyperplane can be freely oriented, and even if one considers only effective configurations of instances on opposite sides of the hyperplane for any larger subset of training instances, the number of combinations is huge. And we should remember that usually, for a single situation, where, for example, a linear separation is possible, a hyperplane location can vary widely. Moreover, a hyperplane's definition involves up to all the features, but from an interpretation perspective, as few features as possible are desired. This means that when searching of oblique splits, a preferable method should be equipped with an embedded feature selection.

As pointed out in Chap. 2, evolutionary techniques have been successfully applied to search for an oblique test but mainly in a top-down induction. In the global approach presented in this book, simultaneous searches for all the needed hyperplanes in non-terminal nodes can rather naturally be built in our single evolutionary algorithm. This necessitates a few extensions of the univariate tree version presented so far. Especially, additional variants of genetic operators need to be proposed, and the complexity term of the fitness function should be revised.

It may seem that replacing inequality tests by oblique splits with embedded feature selection is a very universal and always applicable solution, but experimental results and more deep reflection reveal that it is not really so. In most practical applications of predictive data mining, users want to understand, or at least be able to somehow

© Springer Nature Switzerland AG 2019
M. Kretowski, *Evolutionary Decision Trees in Large-Scale Data Mining*,
Studies in Big Data 59, https://doi.org/10.1007/978-3-030-21851-5_5

control/monitor, how recommendations are obtained. Often, very complex black-box approaches are ignored in favor of maybe slightly worse, but interpretable predictors. This means that as-simple-as-possible tests and models are preferred when constructing a decision tree. It is possible to search for a univariate test within a hyperplane representation, but obviously, it will be significantly less efficient. It becomes evident that heterogeneous trees with unrestricted test representations are necessary.

An interesting issue arises here regarding how to choose the most adequate representation in every node. One of the first attempts among top-down inducers is the CART system [2], where linear splits with all non-nominal features are allowed. However, univariate tests are preferred, so oblique spits are hardly observed. In [3], Brodley investigates a recursive automatic bias selection in a tree-structured hybrid classifier construction. In non-terminal nodes, three types of splits are considered: univariate tests, linear machines, and tests based on K-nearest neighbour. The proposed heuristic selection algorithm takes into account both the characteristics of the training dataset and various goodness of split measures. Omnivariate decision trees, where binary splits can be univariate, linear, or nonlinear, are introduced in [4]. The authors use repeated cross-validations and comparative statistical tests on the accuracy to automatically match the complexity of nodes with subproblems defined by the data reaching these nodes. If there are no statistically significant differences, simpler tests are chosen. In [5], the computationally demanding cross-validation procedure is eliminated, and a new classifiability measure related to the Bayes error and the boundary complexity is proposed instead. Yildiz [6] proposes using structural risk minimization (SRM) to select the node types during omnivariate tree top-down induction and confronts the SRM-based selection with AIC, BIC, and cross-validation approaches.

All these aforementioned methods are embedded into a greedy step-wise tree construction, and any selection is only locally justified. Similarly, like in the case of homogenous trees, such local choices can be suboptimal from a global point of view because they can overlook important interactions between various splits. Fortunately, evolutionary methods seem to be perfectly suited to a global search for proper test representations in all non-terminal tree nodes as well. This approach requires the admission of the various test representations and their differentiation and mutual replacement. A complexity-aware inductive bias needs to be defined on a tree level through an appropriate fitness function. The first evolutionary-based system for a mixed decision tree induction is presented in [7]. It can evolve classification trees with inequality and oblique tests by means of a fine-grained parallel algorithm.

In [8], Magana-Mora and Bajic propose a kind of meta-system approach that uses a tree structure to combine various classifiers (namely artificial neural network, random forest, multinomial logistic regression, and decision trees) from Weka [9]. In the chromosome of a parallel genetic algorithm, their *OmniGA* system encodes the parameters of different models that are embedded in all internal nodes. The validation dataset (15%) extracted from the training data is used to estimate accuracy, and it serves a fitness function. Moreover, certain deep learning ideas are adopted; each successive non-terminal node uses as its new features the predicted probabilities for each instance in the parent node. In addition, ensemble learning methods are also coupled. It should be clearly stated that the result of this interesting system is really far from a typical decision tree.

Also, the same evolutionary approach can be applied to autonomically choose the model type in leaves for regression problems. In the GDT system, it is possible to simultaneously search for test representations in internal nodes and model representations in leaves in a single evolutionary algorithm.

5.1 Representation and Initialization

The representation of oblique or mixed trees is analogical, as in the case of univariate trees, but obviously, a new test type is introduced. For mixed trees, oblique tests are added as a third possibility, whereas for oblique ones, only oblique tests are considered. Each hyperplane is represented as a fixed size $M + 1$—dimensional table of floating-point numbers corresponding to M weights w_j and a threshold θ. Here, zeroing a value of w_j is equivalent to an elimination of the j feature. It is also assumed that only continuous-valued (possibly normalized) features or pre-encoded nominal features are taken into account in the creation of an oblique test.

For the initialization of a population of oblique trees, a dipolar top-down induction is slightly adapted. When a dipole (k, l) is randomly chosen in a node (a mixed dipole for classification or a long dipole for regression), a position of the splitting hyperplane $H_{k,l}(\mathbf{w}, \theta)$ is chosen in such a way that the dipole ends are situated on opposite sides of the hyperplane. It is equivalent to:

$$(\langle \mathbf{w}, \mathbf{x}_k \rangle - \theta) * (\langle \mathbf{w}, \mathbf{x}_l \rangle - \theta) < 0. \tag{5.1}$$

The hyperplane can be defined as follows:

$$\mathbf{w} = \mathbf{x}_k - \mathbf{x}_l, \tag{5.2}$$

and

$$\theta = \sigma * \langle \mathbf{w}, \mathbf{x}_k \rangle + (1 - \sigma) * \langle \mathbf{w}, \mathbf{x}_l \rangle, \tag{5.3}$$

where $\sigma \in (0, 1)$ is a randomly drawn coefficient that determines the distances of the dividing hyperplane to the dipole ends. The hyperplane $H_{k,l}(\mathbf{w}, \theta)$ is perpendicular to a segment connecting the (k, l) dipole ends. In Fig. 5.1, a construction of the hyperplane is depicted.

In decision trees, the number of available training instances for constructing a split decreases with each tree level. In the lower parts of a tree, nodes with a small number of instances are possible. This can result in an "under-fitting" problem [10], where finding a perfect split is very easy, but the split can be rather incidental and can lead to performance degradation. To prevent such a situation, it is assumed that the number of training instances should be meaningfully higher than the number of features used to search for a splitting hyperplane. In the GDT system, the maximal

Fig. 5.1 A splitting
hyperplane construction
based on a randomly chosen
long dipole

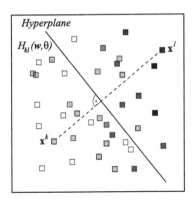

number of features in a test is restricted using the number of available instances
in a node (default value is five instances for each non-zero feature weight). This
restriction needs to be applied during the aforementioned dipolar initialisation. When
the number of non-zero weights exceeds the imposed limit, in a step-wise manner, a
feature to be zeroed is randomly selected, and the threshold is accordingly modified.
In this case, feature dropping is stopped when the limit is finally reached. A very
similar limitation involves the training instances in a leaf and its multivariate linear
model.

For oblique and mixed trees, we decide to abandon the possible memetic initialization variants (e.g., based on SVM [11] or dipolar function minimization [12])
mainly because of computational complexity reasons.

The initialization of a mixed tree population is a little bit more complicated, and it
depends on the type of a predictive task. For classification problems, one half of the
population is initialized with univariate trees and the second half with oblique ones
[13]. In the case of regression problems, when the models in leaves are accepted, we
decided [14] to distinguish even five different groups of equal size:

- Univariate regression trees
- Oblique regression trees
- Univariate model trees
- Oblique model trees
- Mixed trees with both univariate and oblique tests and regression and model leaves.
  At each step of recursive partitioning, a test type or a leaf type is randomly selected.

In this way the initial population is really diverse, and various tree types can compete
for a win.

5.2   Genetic Operators

Introducing oblique tests forces the development of additional mutation variants or
the modification of existing mutation variants. Moreover, some minor changes are
needed in the crossover. Let us start with the modification of an oblique split. It

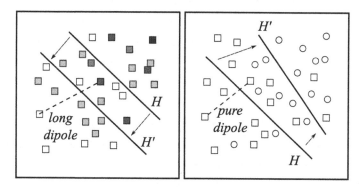

Fig. 5.2 Examples of dipolar shifts

can be executed either by a simple modification of a randomly chosen hyperplane parameter (one of the weights or the threshold) or by a hyper-plane shift based on a randomly chosen dipole. In the dipolar operator, there are two possible situations depending on if one wants to cut a dipole or to avoid cutting it. In the first case, a not divided mixed (for classification) or long (for regression) dipole is searched for. The second case applies only to classification, and a divided pure dipole is chosen. In both cases, the new location of a splitting hyperplane is obtained by modifying just one of its parameters, which is randomly chosen. In Fig. 5.2, examples of dipolar shifts are presented.

In oblique trees, the random modification of hyperplane parameters is not enough to cause an effective simplification of a test because zeroing a parameter is very unlikely. To overcome this problem, an additional variant of oblique test mutation is introduced, one that randomly selects a parameter and assigns zero to it. By increasing the probability of this variant selection, a user can influence the inductive bias.

In the case of mixed trees, where different representations of tests (and models) are allowed, it is necessary to ensure the high diversity of the various building blocks during the evolution. This can be realized by embedding into the mutation two mechanisms: a representation of any new element (test or model) should be randomly selected and changing the representation of an existing element should be possible. The first mechanism is applied in all these mutation variants when a new test is searched, for example, when a leaf is converted to an internal node. In such a situation, univariate or oblique tests can be constructed with equal probability. In the case of univariate tests, it is the type of the randomly selected feature that determines the test representation. When a new test is constructed to replace the existing one, the test type is also flipped with a given probability: an oblique test could be replaced with a univariate one and vice-versa. Similarly, when a leaf is considered, its prediction can be calculated based on a multivariate linear model, or the target feature average can be simply returned.

A test representation change is not only possible through a new test creation, but also when a test is modified [13]. It should be noticed that any inequality test $A_i < th$

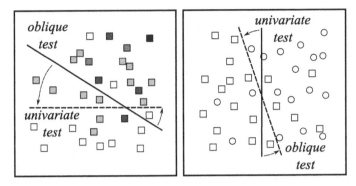

Fig. 5.3 Switching test representations among univariate and oblique tests

can be expressed as an axis-parallel hyperplane $H(\mathbf{w}, \theta)$, where $\theta = th$, $w_i = 1.0$ and where all other weights are equal to zero ($w_j = 0$, for $i \neq j$). Hence, to modify a univariate test to related oblique one, first, a test representation is switched, and then, one zero weight is randomly disturbed. In this way, the axis-parallel split is slightly tilted. An inverse operation is also possible when an oblique split is aligned. This is realized by zeroing all weights except the highest one, and then, switching the test representation into a univariate one is possible with a threshold $th = \frac{\theta}{w_i}$, where w_i is the non-zero weight. In Fig. 5.3, the concept of switching test representations is sketched.

5.3 Fitness Function

For oblique and mixed tree induction, modifications of the fitness function are necessary only in the $Compl(T)$ term. It is obvious that oblique splits should be treated as more complex than univariate ones. Typically, this refers to the number of involved features in a test that can be used as a measure of complexity of the test. For an oblique split, this corresponds to the number of non-zero weights in a hyperplane. A similar idea was employed in the previous chapter to characterize the complexity of a regression model in a leaf. Hence, when both oblique splits and multivariate models are permitted, the complexity of a tree T can be simply defined as:

$$Compl(T) = S(T) + MS(T) + TS(T), \tag{5.4}$$

where $TS(T)$ stands for the overall complexity of the tests in a tree T and is calculated as the sum of all test complexity in it. A slightly more general version of the complexity could be defined as follows:

$$Compl(T) = S(T) + \beta_M \cdot MS(T) + \beta_T \cdot TS(T), \tag{5.5}$$

Fig. 5.4 An example of the
hyperplane centering

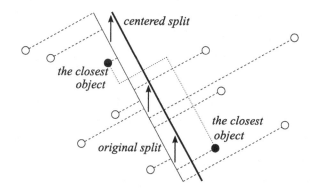

where β_M and β_T are the relative importance coefficients of the model and test complexity and could be provided by a user. Indeed, they can be modified to incorporate the user's preferences. For a univariate classification tree T, both $MS(T) = 0$ and $TS(T) = 0$, so the complexity is reduced to the tree size $S(T)$. In [14], we investigate a mixed model tree induction where the number of nodes is twice as important than the number of features in the tests and models ($\beta_M = \beta_T = 0.5$).

5.4 Post-evolution Processing

It is observed in [15] that enlarging the margins[1] in oblique decision trees could improve classification accuracy. A simple mechanism, called hyperplane centering, inspired by this observation is introduced [13] in the GDT system. When the evolutionary induction is finished, every oblique test in the best-found tree is analyzed, and it is possibly centered. On the opposite sides of the splitting hyperplane, the closest objects are determined, and if these objects have different values for the dependent variable, the hyperplane is shifted to the center by modifying a threshold; as a result, the distances are aligned. It should be noted that in terms of the discussed fitness functions, this post-processing does not change the fitness value. In Fig. 5.4, the split centering mechanism is depicted.

However, hyperplane centering in oblique splits is analogous to the threshold shifting in univariate tests.

5.5 Experimental Results

Two sets of experimental works are presented. First, the influence of the representation (which encompasses both the tests in internal nodes and the models in terminal nodes) on the quality of the generated trees is studied. It is also verified whether

---

[1]A margin is defined as the distance between the decision boundary and the closest training objects.

the evolutionary inducer can autonomously choose an adequate representation both globally and locally. The second set tries to confront the mixed version of the GDT inducer with classical counterparts on regression problems.

5.5.1   The Role of Tree Representation in Regression Problems

To demonstrate the impact of the proper representation of tests and predictions in leaves on decision tree performance, four homogenous and one heterogeneous variants of the GDT system are considered. In the case of homogenous inducers, the representation is fixed and cannot be switched during the evolution. Initialization, genetic operators, and the fitness function are correspondingly restricted. The following variants are considered:

- urGDT—only univariate tests are permitted, and leaves are associated with single constant predictions (averages);
- umGDT—it uses univariate tests in internal nodes and linear models in leaves;
- orGDT—only oblique tests are enabled, but the predictions in leaves are based on constant values;
- omGDT—it uses only oblique tests in internal nodes and linear models in leaves;
- mixGDT—both axis-parallel and oblique tests are possible; similarly, both simple and more advanced models can be used in leaves; the algorithm tries to self-adapt the tree's representation to the analyzed dataset. The resulting representation can be heterogeneous.

Five artificial datasets with analytically defined borders and predictions are analysed. They are modifications [16] of the *Armchair* dataset introduced in the previous chapter, and they require at least three internal nodes and four leaves. There are 1000 instances with two predictors and the target in each dataset. The datasets are divided into the training (66.7%) and testing (33.3%) parts. Each dataset fits a different tree representation: univariate regression (*UR*), univariate model (*UM*), oblique regression (*OR*), oblique model (*OM*), and mixed (*Mix*). Two of the most complex dataset variants are depicted in Fig. 5.5.

Because the generated datasets are not big, each algorithm is run fifty times (with default values for the parameters), and the averages are calculated. In Fig. 5.6, the obtained RMSE, tree size (expressed as the number of leaves), number of features in tests,[2] and number of features in models[3] are provided. All these resulting characteristics should be analyzed together. The optimal number of the features in tests for datasets with oblique tests are equal six for *OR* and *OM*, and four for *Mix*, whereas

---

[2]In univariate trees, the number of features in the tests is simply equal to the number of internal nodes.

[3]For regression trees, where only constant predictions are permitted in leaves, this measure is equal to zero.

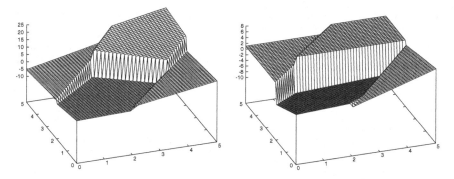

Fig. 5.5 Examples of the *Armchair* dataset variants (*OM* on the left and *Mix* on the right)

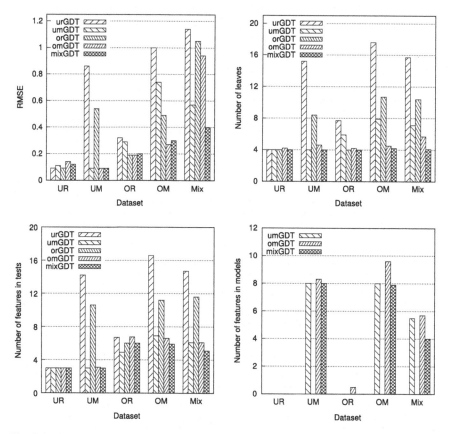

Fig. 5.6 The results obtained by global induction with five distinct representations. Each graph presents a different characteristic. Certain bars are omitted when not applicable

the optimal number of features in the models are eight for *UM* and *OM* and four for *Mix*.

For the simplest dataset (*UR*), all inducers find perfect (or almost perfect) solutions with low errors. Only univariate splits in the internal nodes are used, leaving no linear models in leaves. This dataset may seem to be the most difficult for the omGDT variant, but because a univariate split is a special case of an oblique one and a constant prediction is also a special case of a linear regression model, the necessary representation simplifications are not so problematic.

For the *UM* dataset, both regression variants (urGDT and orGDT) have some expected problems, as the resulting trees are overgrown (16 or 8 leaves instead of 4), and the prediction errors are worse. Furthermore, the number of features in the tests also grows. Concerning the models in leaves, all inducer variants, which can search for linear models, managed to find appropriate models.

It should be noted that the *OR* dataset turns out not to be very difficult for any variant of the GDT inducer. The prediction errors are slightly worse for univariate trees (urGDT and umGDT), and the tree sizes are a little bit larger than optimal.

Regarding the *OM* dataset, only two variants with the most complex representation (omGDT and mixGDT) find the best solution. For the simplest inducer (urGDT), the estimated prediction error is three times higher, and the tree size is four times bigger. The orGDT variant is only slightly better in terms of the prediction error, whereas the umGDT variant obtains smaller trees.

Concerning the last dataset, the obtained results confirm that only an inducer with a heterogeneous representation (mixGDT) find the desired solution without problems. As expected, the omGDT variant is rather close, but it seems that the assumed number of generations is too small. The tree structure is close to optimal, but the prediction error is not. Interestingly, the umGDT version performed relatively well, which can be explained by a compensation of not optimal splits by more numerous but better models.

In Fig. 5.7, the convergence of evolutionary processes for five variants of the GDT system is studied on the simple *UR* dataset. Both the prediction errors and tree size are presented for the best individuals through the first 3000 generations.

It is rather clear that with a more complex representation, the search space is enlarged, and this naturally leads to a slower convergence of the search process. The oblique model variant (omGDT), which searches simultaneously for an adequate tree structure and all oblique splits and linear models in leaves, is the slowest but finally finds a good solution. Interestingly, the mixed variant with a more flexible representation can quickly detect the most convenient representation for the problem studied, and this results in a noticeably faster convergence. Concerning the tree sizes, the observed initial tree sizes are very diverse (for mixGDT, the trees are overgrown, whereas for omGDT, they are too simple), and the sizes evolve to finally stabilize on the desired value. Only for the omGDT variant are more than 3000 generations needed to find the optimal tree size.

In addition, the average time of an iteration for omGDT is almost two times longer than for mixGDT (0.0043–0.0024 s), which additionally emphasizes the advantage

Fig. 5.7 The performance of the best individuals for various tree representations on the *UR* dataset: RMSE (top) and number of nodes (bottom)

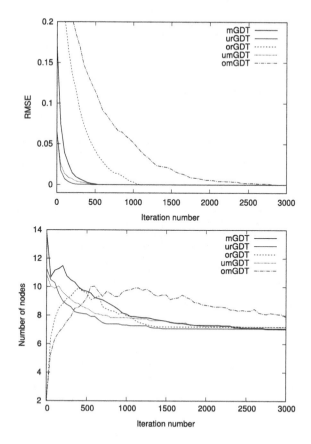

of mixed trees. On the other hand, for the simplest univariate regression, the average generation time is obviously shorter (0.0013 s).

5.5.2 *Mixed GDT for Regression*

The performance of a mixGDT variant is compared with well-known decision tree inducers on regression problems. From the benchmark datasets (available on the WEKA webpage) provided by Torgo [17], this time, only the results obtained on all larger datasets (with more than 10,000 instances) are presented. The dataset characteristics are given in Table 5.1. If necessary, the nominal features from the original datasets are converted into binary features, and the missing values are replaced by the corresponding mean values. The datasets are also divided into two equal-size parts used for training and testing.

In Table 5.2, besides the results obtained by decision tree inducers like RT and M5, the results obtained by the well-known non-single tree predictors are also included

Table 5.1 Characteristics of the selected datasets for the mixed tree evaluation

Dataset	Instances	Features
2dplanes	40768	10
Ailerons	13750	40
Cal housing	20640	8
Fried	40768	10
House 16H	22784	16
House 8L	22784	8
Mv	40768	10
Pole	15000	48

Table 5.2 The prediction error (RMSE) estimated for five inducers. The results for some datasets are multiplied by the same value (by 10^4 for *Ailerons* and by 10^{-3} for *Cal housing, House 16H, and House 8L*) to facilitate a visual inspection

Dataset	mixGDT	RT	M5	SVM	Bagging
2dplanes	1.0	1.05	1.0	2.38	1.04
Ailerons	1.6	2.0	1.6	1.7	1.8
Cal housing	7.4	9.7	13.3	7.4	7.8
Fried	1.07	1.92	1.56	2.66	1.50
House 16H	3.9	3.9	3.6	4.6	3.4
House 8L	3.3	3.5	3.2	4.6	3.1
Mv	0.09	0.35	0.21	5.3	0.20
Pole	7.08	8.96	6.87	30.7	6.37

for reference (SVM SMOreg [18] that implements the support vector machine for regression and Bagging [19] as an example of an ensemble of regression trees). All competing inducers are implemented in Weka, and they are run with a default set of parameter values. The presented results do not encompass the complexity of the obtained predictors because a direct comparison of a single tree size with an ensemble or SVM-structure size could be very misleading.

Here, mixGDT is a really competitive inducer even if compared with currently appreciated approaches like ensembles or SVM. For these eight datasets, the global inducer gains the smallest error in five cases. In other cases, it is just behind the best solution, and it never visibly fails, as seen for SVM (*Pole* and *Mv* datastets) or M5 (*Cal housing*). A more detailed comparison of the mixed version and other variants of the GDT system with other systems on a bigger set of problems can be found in [14], where a statistical analysis of the results is also presented.

References

1. Murthy S (1998) Data Min Knowl Discov 2:345–389
2. Breiman L, Friedman J, Olshen R, Stone C (1984) Classification and regression trees. Wadsworth and Brooks, Monterey
3. Brodley C (1995) Mach Learn 20(1–2):63–94
4. Yildiz O, Alpaydin E (2001) IEEE Trans Neural Netw 12(6):1539–1546
5. Li Y, Dong M, Kothari R (2005) IEEE Trans Neural Netw 16(6):1547–1560
6. Yildiz O (2011) Inf Sci 181(23):5214–5226
7. Llora X, Wilson S (2004) Mixed decision trees: minimizing knowledge representation bias in LCS. In: Proceedings of GECCO'04. Lecture notes in computer science, vol 3103, pp 797–809
8. Magana-Mora A, Bajic V (2017) Sci Rep 7:3898
9. Frank E, Hall M, Witten I (2016) The WEKA workbench. Online appendix for "Data mining: practical machine learning tools and technique", 4th edn. Morgan Kaufmann, Burlington
10. Duda O, Heart P, Stork D (2001) Pattern classification, 2nd edn. Wiley, New York
11. Kumar M, Gopal M (2010) Pattern Recognit 43(12):3977–3987
12. Bobrowski L, Kretowski M (2000) Induction of multivariate decision trees by using dipolar criteria. In: Proceedings of PKDD'00. Lecture notes in computer science, vol 1910, pp 331–336
13. Kretowski M, Grzes M (2007) Int J Data Wareh Min 3(4):68–82
14. Czajkowski M, Kretowski M (2016) Appl Soft Comput 48:458–475
15. Bennett K, Cristianini N, Shave-Taylor J, Wu D (2000) Mach Learn 41:295–313
16. Czajkowski M, Czerwonka M, Kretowski M (2013) Cost-sensitive extensions for global model trees. Application in loan charge-off forecasting. In: Advances in systems science. Advances in intelligent systems and computing, vol 240, pp 315–324
17. Torgo L (2018) Regression datasets repository. http://www.dcc.fc.up.pt/~ltorgo/Regressio/DataSets.html
18. Shevade S, Keerthi S, Bhattacharyya C, Murthy K (2000) IEEE Trans Neural Netw 11(5):1188–1193
19. Breiman L (1996) Mach Learn 24(2):123–140

Part III
Extensions

Chapter 6
Cost-Sensitive Tree Induction

In most typical data mining research, there is the assumption that all prediction errors are equally important. Similarly, all features are usually treated as equal whenever available. In real-world applications in medicine or business, such an idealization is not always possible. Often, a cost-sensitive prediction is desirable, and it should account for various cost types, starting with the asymmetric costs (losses) associated with predictive errors.

In medical decision making depending on the type of error, the consequences of the wrong diagnosis can have a significantly different impact. Misclassifying an ill patient as a healthy one is usually dangerous for the patient and can result in severe complications. An inverse situation, when a healthy person is mistakenly classified as a sick person is stressful. But before any intervention, additional examinations are prescribed, and typically, the matter is clarified. Similarly, for a banking regulatory authority, incorrectly classifying a failing bank as a sound one is much more harmful than raising a possibly false alarm on a healthy bank [1]. In both situations, considering the varying costs of misclassification regarding whether an instance is a false negative (a positive example classified as negative) or a false positive (a negative example classified as positive) enables constructing reasonable and effective classifiers.

In regression problems, the cost of misprediction is naturally related to the size of an error, but the asymmetry of the losses is also common. For example, when speculating on the stock exchange, investors analyze the future gains and losses, but typically, they pay more attention to losses. According to [2], potential gains should be approximately twice as large to offset potential losses. As a result, investors tend to realize their gains more often because they sell winning stocks more readily. In bank reserve forecasting, underpredicting future loan charge-offs by a certain amount is riskier than overprediction by the same amount [3]. If a bank overpredicts its future loan charge-off, it will need to maintain extra funds in its reserves and thus suffer reduced earnings. Possibly, it will receive a lower credit score from financial analysts. However, underpredicting a loan charge-off could have more serious consequences

© Springer Nature Switzerland AG 2019
M. Kretowski, *Evolutionary Decision Trees in Large-Scale Data Mining*,
Studies in Big Data 59, https://doi.org/10.1007/978-3-030-21851-5_6

because the bank is not adequately prepared for its future loan losses. If the bank does not maintain sufficient provisions, it will not only face the wrath of investors and regulators, but also experience a large credit rating downturn.

There are many types of cost that can be identified in inductive learning [4]. Turney proposes a taxonomy of types of cost with nine major elements (e.g., misclassification costs, test costs, teacher costs, computation costs, or intervention costs) and many possible subtypes that take into account other aspects (e.g., constant or conditional, dynamic or static). The costs are not necessarily expressed in the same units, so for combining them, monetary units are typically used as a unification platform. It is also possible to propose a heterogeneous cost-sensitive decision tree induction [5]. Cost-sensitive learning is now an active research area, with most effort concentrated on incorporating varied misclassification costs and/or test costs in classification problems. Decision trees are very often applied as a handy tool in cost-sensitive classification [6] but rather occasionally in regression.

There are three main approaches for cost-sensitive decision trees. In the first group, the existing cost-free induction methods are converted or extended. This usually rests on modifying a splitting criterion or a post-pruning procedure. One of the first attempts to incorporate misclassification costs is the CART system [7]. The class, prior probabilities used in the splitting criterion are altered, and a cost-based measure is applied in the tree pruning. A lot of cost-sensitive pruning methods have been proposed. In [8], the misclassification costs are integrated into several well-known pruning techniques, whereas in [9], the authors study a pruning variant based on the Laplace correction. On the other hand, cost-sensitive pre-pruning strategies have also been investigated [10]. However, it should be emphasized that stand-alone pruning procedures have only a limited capability to change the tree structures created by error-based inducers.

In the EG2 system [11], Nunez modifies an information theoretic measure for feature selection by introducing a feature cost. A user can vary the bias extent through a parameter. Because the source code of the C4.5 [12] system is downloadable, it is the system that has been extended in many works, so only a few can be mentioned. In [13], it is modified using instance-weighting, but the method requires converting a cost matrix into a cost vector, which can result in poor performance in multi-class problems. In [14], two adaptive and heuristic mechanisms (for selecting the cut-off points and removing features) are introduced to improve the efficiency of larger datasets (up to twenty thousand objects and/or features). The algorithm minimizes both the test and misclassification costs. In their next paper [15], the authors propose another heuristic function to evaluate the features in a node.

The second large group of approaches encompasses the general methods for making an arbitrary predictor (typically classifier) cost-sensitive, which can be applied to decision trees. Only a few representative examples are recalled here. One of the earliest (and extensively applied) approaches is MetaCost [16], proposed by Domingos. The algorithm is based on wrapping a bagging-like meta-learning stage around any error-based classifier. Zadrozny et al. [17] introduce cost-proportionate rejection sampling and ensemble aggregation. Then, they investigate [18] iterative weighting and gradient boosting in multi-class problems. In [3], Bansal et al. propose a post hoc tuning method for regression, which minimizes the average misprediction cost under an asymmetric cost structure. The method uses a hill-climbing algorithm to find an adjustment that is added to the prediction of the regular regression model. In their next paper [19], the authors extend the approach by searching (with iterative hill-climbing) for a polynomial function, which then adjusts the model.

The last group of approaches consists of developing dedicated cost-sensitive decision tree inducers. Zhang et al. [20] propose an algorithm that minimizes the sum of the misclassification and test costs. This approach is based on a new splitting criterion (total cost) for nominal attributes and two-class problems. In [21], a slightly modified cost model inspired by real-life business applications, where the costs because of misclassification vary between examples, is investigated. The authors propose an example-dependent cost-sensitive decision tree induction by incorporating the different example-dependent costs into a new cost-based impurity measure and hence produce new cost-based pruning criteria. Recently, Lomax and Vadera [22] have presented an algorithm based on a multi-armed bandit game that searches for a trade-off between decisions based on accuracy and decisions based on costs. Interestingly, the algorithm selects the attributes during decision tree induction by using a look-ahead methodology to explore the potential attributes and exploit the attributes that maximize the reward.

Concerning an evolutionary computation application to cost-sensitive decision tree learning, two works should be mentioned. Turney proposes Inexpensive Classification with Expensive Tests (ICET) [23], which evolves a population of biases for the modified C4.5 inducer with the EG2 cost function by means of a standard genetic algorithm. Both the misclassification and feature costs are considered. Interestingly, the features can be grouped and share the cost, as is often observed in a medical diagnosis. The fitness is measured by an average cost of classification estimated on a sub-testing set (a randomly selected half of a dataset). A global system for searching of tree based-structures with logical formulas in non-terminal nodes based on genetic programming is introduced in [24]. It uses a class-related object weighting and constrained fitness in solving two classes problems.

In this chapter, two cost-sensitive extensions of a global decision tree induction will be presented. First, a cost-sensitive classification is studied, and both the misclassification costs and test costs are incorporated into the tree evolution. Then, asymmetric loss functions are considered, and cost-sensitive model trees can be generated. It is important to underline that the necessary changes in the GDT system are very limited, revealing the elasticity of evolutionary induction.

6.1 Cost-Sensitive Classification

A transformation of the original GDT system into a cost-sensitive learner requires only a few changes. Mainly, the definition of the $Err(T)$ component from the weighted formula fitness function must be revised to incorporate the test costs and misclassification costs. In addition, a mechanism for assigning a class label to a leaf should be slightly adapted, and a class label is chosen to minimize the misclassification cost in a leaf. And finally, the uniform probability of choosing a feature for a new test could be skewed to prefer less costly features. This is realized by the sorting features according to their cost and then assigning their probabilities in the same way as in the ranking linear selection. If any feature has been already used in prior tests, its cost is reduced to zero. This can be explained by the fact that obtaining a feature value is costly (e.g., performing a blood test in a laboratory) but not reusing this value in a tree.

Let a feature cost corresponding to a feature A_i be denoted as C_i and $Cost(d_i, d_j) \geq 0$ be a cost of misclassifying an object from the class d_j as belonging to the class d_i. It can be assumed that the cost of a correct decision is equal to zero $(Cost(d_i, d_i) = 0)$ for all classes.

The performance of a typical classification tree is judged by a reclassification error, whereas a cost-sensitive tree should be assessed by an average cost, which is the sum of an average misclassification cost and an average test cost.[1] This suggests a direct substitution of the error by the cost in the fitness function. In general, an expected cost is not limited because it is related to external cost values, but for a given dataset, cost matrix, and feature costs, it is possible to calculate the maximum misclassification and test costs. As a result, a normalized cost can be easily obtained by dividing an observed cost by the maximum, and thus, such a normalized value can fit into the [0, 1] range.

6.1.1 Misclassification Cost

Hence, an average misclassification cost $MCost(T)$ [25] of a given tree T is estimated on a training dataset L composed of M instances $\mathbf{x}_i$:

$$MCost(T) = \frac{1}{M} \cdot \sum_{i=1}^{M} Cost(T(\mathbf{x}_i), d(\mathbf{x}_i)). \tag{6.1}$$

The maximal misclassification cost $MaxMC$ for a given dataset L and a cost matrix can be calculated according to the following:

$$MaxMC = \frac{1}{M} \cdot \sum_{d_k \in D} |D_k| \cdot \max_{i \neq k} Cost(d_i, d_k). \tag{6.2}$$

This corresponds to a situation where all the objects are incorrectly classified and the wrong predictions are, at each time, the most expensive. The normalized (average) misclassification cost obtained by dividing $MCost(T)$ by $MaxMC$ is equal to zero for the perfect prediction and one in the worst possible case:

---

[1]It is assumed that both costs use the same measurement units.

$$NormMC(T) = \frac{MCost(T)}{MaxMC}. \tag{6.3}$$

6.1.2 Test Cost

Similarly, a normalized average test cost of a tree T generated from a training dataset L can be obtained.

Let $A(T, \mathbf{x}_i)$ denote a set of all the features from the tests on a path from the root node of a tree T to a leaf reached by an object $\mathbf{x}_i$. Hence, the cost of all tests $TC(T, \mathbf{x}_i)$ necessary to classify an object $\mathbf{x}_i$ by a tree T is equal to:

$$TC(T, \mathbf{x}_i) = \sum_{A_i \in A(T, \mathbf{x}_i)} C_i. \tag{6.4}$$

It should be noted that the renewed use of any feature in the subsequent tests does not increase $TC(T, \mathbf{x}_i)$. Then, an average test cost $TCost(T)$ can be derived as follows:

$$TCost(T) = \frac{1}{M} \cdot \sum_{i=1}^{M} TC(T, \mathbf{x}_i). \tag{6.5}$$

The maximal cost of tests for any instance from the training dataset L and fixed costs of the features can be obtained as follows:

$$MaxTC = \sum_{i=1}^{N} C_i, \tag{6.6}$$

where all the features are indispensable for a prediction. As expected, a normalized average test cost can be calculated by dividing $TCost(T)$ by $MaxTC$:

$$NormTC(T) = \frac{TCost(T)}{MaxTC}. \tag{6.7}$$

For a degenerated tree composed of a single leaf, it is equal to zero, and for an overgrown tree with all the features in tests on every path from a root to any leaf, it is equal to one.

6.1.3 Misclassification and Test Costs

If both costs need to be considered and it is assumed that they are equally important, a simple sum of the normalized costs can be applied in the fitness function:

$$Err(T) = NormMC(T) + NormTC(T), \tag{6.8}$$

but it doubles the range of $Err(T)$. This can be easily resolved by the following formula [26]:

$$Err(T) = \frac{MCost(T) + TCost(T)}{MaxMC + MaxTC}. \tag{6.9}$$

6.2   Cost-Sensitive Regression

Typical loss functions such as the absolute error or the squared error are symmetric, and they dominate in the statistics and data analysis. Granger [27] points out serious problems when choosing the right loss function, which is assumed to be symmetric, but actually, the loss functions are often asymmetric. The $LinEx$ loss function, introduced by Varian in [28], is one of the first asymmetric cost functions in prediction and has become a popular alternative to least squares procedures. $LinEx$ is approximately exponential on one side and linear on the other. Later, further asymmetric loss functions (see Fig. 6.1) are proposed [29]: $LinLin$ (asymmetric linear) and $QuadQuad$ (asymmetric quadratic).

The transformation of the global model tree induction into a cost-sensitive variant requires at least a few modifications. First, the fitness function should be able to consider various misprediction costs. In addition, because not all differentiation operators are totally random, it could be useful to introduce some cost-aware variants. So we propose a post-genetic operation model tuning and an additional mutation variant. Finally, the misprediction costs are taken into account when the ranking lists for applying genetic operators are created. More in-depth changes are possible concerning a cost-sensitive initialization or more advanced memetic extensions, but it seems that even rather limited modifications are enough to obtain competitive results.

Fig. 6.1 Asymmetric cost functions

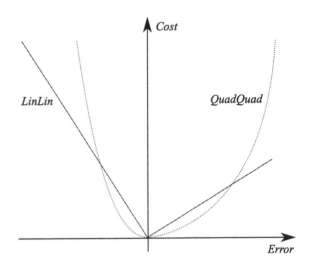

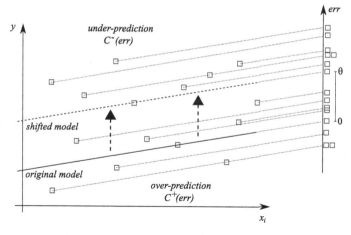

Fig. 6.2 An example of a cost-sensitive model adjustment

Incorporating the misprediction asymmetric costs into the BIC-based fitness function can be realized by replacing the residual sum of squares $RSS(T)$ by the average misprediction cost (AMC) [3]. $AMC(T)$ can be defined as:

$$AMC(T) = \frac{1}{M} \cdot \sum_{i=1}^{M} CS(tf_i - T(\mathbf{x}_i)),\qquad(6.10)$$

where $CS(err)$ is a function that defines a cost corresponding to a prediction error err. Such a small modification is enough to enforce the induction of cost-sensitive model trees. We also investigate possible modifications of the second term devoted to a tree's complexity. In [30], we double a penalty for overparametrization, and the experimental results show that it reduces the tree size and improves the generalizability.

The post-genetic operation model adjustment is inspired by the post hoc tuning method proposed by Bansal et al. [3]. It consists of shifting a regression model in a leaf toward a smaller AMC by adding a fixed θ value. If the θ is positive, a model is moved toward underpredicted instances, and some of them could change a side relative to the model and become over-predicted. In Fig. 6.2, an example of such a cost-sensitive model adjustment is shown.

For an arbitrary convex cost function (including *LinLin* or *QuadQuad*), an approximated θ adjustment can be found by using a simple algorithm. First, all instances are sorted according to the error values, and this is done separately on each side of the model. Then, the distinct error values are analyzed in a step-wise manner, and the AMCs are calculated. Both directions are theoretically possible, but in most cases, the model will be moved toward underpredicted instances. When a cost starts to increase, the wanted range is found, and the exact θ position is randomly selected as long as two instances are on opposite sides of the model (one instance is underpredicted, and the other is overpredicted). Such a model adjustment is performed after any modification (because of a mutation or crossover) of the models in leaves.

Because a technique for fitting a regression model in a leaf is cost-neutral, we decide to introduce an additional mutation variant [30]. It is applicable to a leaf, and its role is to slightly disorder a standard model construction. Together with mutation variants that add, remove, and replace features used to generate a model, it should extend a search space and enable finding more diverse cost-sensitive models. It randomly selects a single feature and modifies a corresponding weight by a randomly chosen coefficient ϑ ($\vartheta \in [-1, 1]$). Next, this type of modified model is adjusted with the aforementioned technique.

6.3   Experimental Results

The capabilities of the cost-sensitive versions of the GDT system are presented both for classification and regression. In the case of classification trees, first, misclassification costs are only considered, and then, the test costs are introduced. As for model trees, a loan charge-off forecasting problem is analyzed.

6.3.1   Cost-Sensitive Classification Trees

The results obtained for ten exemplary datasets from the UCI Machine Learning Repository [31] are presented to assess the performance of the misclassification cost extension of the GDT system. During the evolutionary induction, only univariate tests are allowed. For comparison purposes, the results of two state-of-the-art cost-sensitive systems are also included: C5.0, which is the commercial version of C4.5 [12], and the MetaCost [16] wrapper (implemented in Weka [32] with J48[2] as an error-based classifier). All inducers are launched with a default set of parameters.

For only two datasets (*german* and *heart*), the misclassification costs are known, and for both of them, the cost ratio is 5:1. For the rest of the datasets, cost matrices are prepared following the approach applied in [13, 33]. In each run of the ten-fold cross-validation, a cost matrix is randomly generated. The off-diagonal elements of the cost matrix are drawn from the uniform distribution over the range [1, 10], and all diagonal elements are always zero. It should be noted that in each single cross-validation run, the same random cost matrix and the same training data splits are used for all algorithms. In Table 6.1, the averages of ten runs are presented [34].

For six of the ten datasets, the cost of the GDT system is lower than two competitors. Moreover, the globally induced decision trees are the smallest for all the datasets. Only for one dataset (*page-blocks*) is the evolutionary inducer worse than C5.0 and MetaCost in terms of cost in favor of a small tree size. However, a sim-

---

[2]J48 represents a Weka implementation of the C4.5 algorithm.

Table 6.1 Misclassification costs and tree sizes obtained with known and generated cost matrices

Dataset	csGDT		C5.0		MetaCost	
	Cost	Size	Cost	Size	Cost	Size
German	0.57	7	0.71	81.4	1.26	31.6
Heart	0.56	10.7	0.56	16.5	1.01	18.0
Australian	0.62	5.4	0.6	12.5	0.63	7.0
Breast-w	0.19	4.4	0.24	11.1	0.29	10.0
Balance-scale	1.23	10.3	1.27	29.8	1.33	34.0
Cars	0.08	3.2	0.05	25.6	0.21	31.0
Glass	1.39	6.4	1.64	20.7	2.40	14.0
Page-blocks	0.24	3.6	0.17	34.6	0.19	41.1
Pima	0.89	4.8	0.93	15.6	1.64	28.2
Vehicle	1.43	11.6	1.53	69.0	1.96	70.9

Table 6.2 Results obtained when both the misclassification and test costs are considered

Dataset	csGDT		ICET-like		EG2-like	
	Cost	Size	Cost	Size	Cost	Size
Bupa (A)	19.13	3	48.42	52.6	19.56	5
Heart (A)	173.21	6.4	371.58	44.8	187.79	20
Hepatitis (A)	11.67	7.1	18.53	5.8	10.06	5
Pima (A)	21.15	4.0	24.53	41.4	23.08	16
Bupa (B)	22.60	1.4	61.86	52.1	22.96	1
Heart (B)	242.72	7.4	478.74	48.3	239.67	20
Hepatitis (B)	17.76	7.4	25.09	5.8	27.51	6
Pima (B)	27.27	5.0	33.34	43.4	25.38	20

ple tuning of the α parameter in the weighted fitness function leads to a substantial improvement, and the cost can be reduced to 0.11 for this dataset. Generally, in [25], an impact of the α parameter on the misclassification cost and the tree complexity is studied in detail.

As for the results obtained with two costs (misclassification and test costs), the global induction is compared with non-native implementations of ICET [23] and EG2 [11] on four medical datasets with known feature costs that are investigated in [23]. For each dataset, a misclassification cost of the less-frequent class is equal to a sum of feature costs, and for more frequent classes, the cost is increased by fifty percent (denoted by A) or 200% (B). In Table 6.2, the obtained costs and tree sizes are presented for two cost ratios.

More detailed experimental results concerning the various cost ratios can be found in [34]. The cost-sensitive version of the GDT system performs very well. It is better than ICET in terms of the costs for all studied configurations and also better that EG2 in five out of eight configurations. The global inducer is also competitive in terms of tree complexity, especially when compared with another evolutionary approach (ICET).

6.3.2 Loan Charge-Off Forecasting

Loan charge-off forecasting is a typical regression problem characterized by asymmetric costs because overprediction is less costly than underprediction. This problem is analyzed by researchers (e.g., [3, 19]) using the data from American financial institutions that are provided by Wharton Research Data Services [35]. From 2004 to 2010, the number of considered banks, depending on the year, varies from 6992 to 8315 (average 7695). In the dataset, only fourteen available features (e.g., *total loans and leases, gross; total assets; goodwill; interest and fee income on loans (IN), risk-weighted assets (RI)* ...) describing quarter by quarter the financial situation of each bank are stored. No macroeconomic and unemployment information is included. The last feature is *charge-offs on allowance for loan and lease losses (CH)*, and these are the values that should be predicted with the data from the previous quarter. The data are preprocessed in the same way as postulated in the aforementioned papers, which means that all records with missing values are eliminated, and the natural logarithm transformation is applied on the target to reduce the skewness. Based on the available data, twenty-six independent training and testing dataset pairs are created, where each induction is performed on a single quarter and an obtained predictor is tested on the next quarter.

The cost-sensitive version of the global model tree induction (denoted as $csGDT$) is compared with the classical regression predictors available in the Weka system [32]: the standard least-squares linear regression (LR), the M5 model tree algorithm (M5), and the back-propagation neural network (NN). All these cost-neutral solutions generated with the default settings are then post-tuned with the *Linear* adjustment [19]. The source code of the tuning method is provided by Huimin Zhao. The experimental results encompass two cost functions (*LinLin* and *QuadQuad*) and three cost ratios for underprediction to overprediction (10:1, 50:1, and 100:1).

In Table 6.3, the average AMC values observed over twenty-six independent testing datasets are presented. For the evolutionary inducer (denoted as $csGDT$), only univariate tests are considered, and twenty repetitions are scheduled, so the averages are presented. The standard deviations observed for $csGDT$ vary from 0.16 to 0.36 for *LinLin* and from 1.30 to 2.91 for *QuadQuad*.

The results obtained by cost-neutral approaches are clearly inferior to cost-sensitive ones, and the differences become huge for the highest cost ratio. Such pronounced differences observed on the real data explicitly show that specialized cost-sensitive methods are indispensable when various costs have to be considered.

Table 6.3 The results of loan charge-off forecasting

Algorithm	LinLin			QuadQuad		
	10:1	50:1	100:1	10:1	50:1	100:1
LR	7.41	34.02	67.27	22.39	103.19	204.19
LR + Linear	3.81	5.23	5.85	11.80	22.64	30.10
M5	7.29	33.23	65.66	139.03	231.73	347.60
M5+Linear	3.88	6.44	7.94	97.21	97.72	107.73
NN	8.16	37.92	75.12	23.74	110.34	180.43
NN+Linear	3.57	5.09	5.80	10.43	18.62	23.42
csGDT	3.05	3.97	4.55	8.75	13.25	16.04

Fig. 6.3 Average misprediction costs of loan charge-off forecasting for three algorithms in 2008 (*LinLin* loss function, cost ratio 10:1)

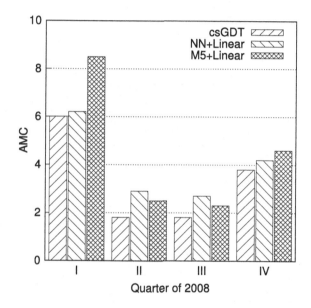

Here, the evolutionary induction of cost-sensitive model trees gives the best results for all cost functions and ratios.

In Fig. 6.3, more detailed results concerning one representative year (2008) and one cost ratio (10:1) are presented. The global approach is confronted with the adjusted M5 and NN. Primarily, the high seasonality [36] of these economic data can be noticed. The first quarter predictions are visibly worse, but it is a rather expected pattern when forecasting financial data. It could be explained by possible changes in bank loan politics after the publication of annual reports and revealing of macroeconomic data. However, it should be underlined that the performance of all methods is consistent and that the global approach predictions are the most accurate for all quarters.

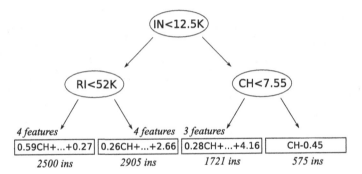

Fig. 6.4 An example of a globally evolved cost-sensitive model tree (the first quarter of 2008, *LinLin* loss function, cost ratio 10:1)

In business analytics, not only is high effectiveness desired, but also the predictor's comprehensibility. Black-box solutions are often abandoned because analysts are interested in methods that can provide new insights into the underlaying process. Fortunately, globally induced model trees are generally small and concise, which is very attractive from an interpretability point of view. In Fig. 6.4, an example of a generated model tree is depicted. It consists of only three non-terminal nodes and four relatively simple models in leaves. The most complicated models are based on four features, and the simplest model involves only one feature. Interestingly, all models use the previous charge-off (CHI), which seems to be reasonable, even for a non-specialist.

References

1. Sarkar S, Sriram R (2001) Manag Sci 47(11):1457–1475
2. Tversky A, Kahneman D (1991) Q J Econ 106:1039–1061
3. Bansal G, Sinha A, Zhao H (2008) J Manag Inf Syst 25(3):315–336
4. Turney P (2000) Types of cost in inductive concept learning. In: Proceedings of ICML'2000 workshop on cost-sensitive learning. Stanford
5. Zhang S (2012) J Syst Softw 85:771–779
6. Lomax S, Vadera S (2013) ACM Comput Surv 45(2):1–35
7. Breiman L, Friedman J, Olshen R, Stone C (1984) Classification and regression trees. Wadsworth and Brooks, Monterey
8. Knoll U, Nakhaeizadeh G, Tausend B (1994) Cost-sensitive pruning of decision trees. In: Proceedings of ECML'94. Lecture notes in computer science, vol 784, pp 383–386
9. Bradford J, Kunz C, Kohavi R, Brunk C, Brodley C (1998) Pruning decision trees with mis-classification costs. In: Proceedings of ECML'98. Springer, pp 131–136
10. Du J, Cai Z, Ling C (2007) Cost-sensitive decision trees with pre-pruning. In: Advances in artificial intelligence. Lecture notes in artificial intelligence, vol 4509, pp 171–179
11. Nunez M (1991) Mach Learn 6(3):231–250
12. Quinlan J (1993) C4.5: programs for machine learning. Morgan Kaufmann, San Francisco
13. Ting K (2002) IEEE Trans Knowl Data Eng 14(3):659–665
14. Li X, Zhao H, Zhu W (2015) Knowl Based Syst 88:24–33

15. Zhao H, Li X (2017) Inf Sci 378:303–316
16. Domingos P (1999) MetaCost: a general method for making classifiers cost-sensitive. In: Proceedings of KDD'99. ACM Press, pp 155–164
17. Zadrozny B, Langford J, Abe N (2003) Cost-sensitive learning by cost-proportionate example weighting. In: Proceedings of ICDM'03. IEEE Press, pp 435–442
18. Abe N, Zadrozny B, Langford J (2004) An iterative method for multi-class cost-sensitive learning. In: KDD'04. ACM Press, pp 3–11
19. Zhao H, Sinha A, Bansal G (2011) Decis Support Syst 51(3):372–383
20. Zhang S, Qin Z, Ling C, Sheng S (2005) IEEE Trans Knowl Data Eng 17(12):1689–1693
21. Bahnsen A, Aouada D, Ottersten B (2015) Expert Syst Appl 42(19):6609–6619
22. Lomax S, Vadera S (2017) Comput J 60(7):941–956
23. Turney P (1995) J Artif Intell Res 2:369–409
24. Li J, Li X, Yao X (2005) Cost-sensitive classification with genetic programming. In: Proceedings of CEC'05. IEEE Press, pp 2114–2121
25. Kretowski M, Grzes M (2007) Evolutionary induction of decision trees for misclassification cost minimization. In: ICANNGA'07. Lecture notes in computer science, vol 4431, pp 1–10
26. Kretowski M, Grzes M (2006) Evolutionary induction of cost-sensitive decision trees. In: ISMIS'06. Lecture notes in artificial intelligence, vol 4203, pp 121–126
27. Granger C (1989) Forecasting in business and economics, 2nd edn. Academic Press, London
28. Varian H (1975) A Bayesian approach to real estate assessment. In: Studies in Bayesian econometrics and statistics. North Holland, pp 195–208
29. Cain M, Janssen C (1995) Ann Inst Stat Math 47(3):401–414
30. Czajkowski M, Czerwonka M, Kretowski M (2015) Decis Support Syst 74:57–66
31. Dua D, Karra Taniskidou E (2017) UCI machine learning repository http://archive.ics.uci.edu/ml. Irvine CA: university of California, School of information and computer science
32. Frank E, Hall M, Witten I (2016) The WEKA workbench. Online appendix for "Data mining: practical machine learning tools and techniques", 4th edn. Morgan Kaufmann, Burington
33. Margineantu D, Dietterich T (2000) Bootstrap methods for the cost-sensitive evaluation of classifiers. In: Proceedings of ICML'2000. Morgan Kaufmann, pp 583–590
34. Kretowski M (2008) Obliczenia ewolucyjne w eksploracji danych. Globalna indukcja drzew decyzyjnych, Wydawnictwo Politechniki Bialostockiej
35. The Wharton school (2018) The university of Pennsylvania, Wharton research data services. http://wrds-web.wharton.upenn.edu/
36. Liu C, Ryan S, Wahlen J (1997) Account Rev 72(1):133–146

Chapter 7
Multi-test Decision Trees for Gene Expression Data

Bioinformatics [1] is now one of the fastest growing and most promising interdisciplinary fields of science, providing the methods and software tools to get insights into information processing in living organisms. This field tries to apply well-established methods from various domains but also develop new algorithms that enable a better understanding of biological data. With the advent of high throughput technologies, it is possible to obtain huge amounts of genomic data in a relatively easy and cheap way. Among the many types of omics data, gene expression data are the most readily analyzed as expression profiles and are perceived as extremely useful in a precise diagnosis (e.g., cancer subtype differentiation) and personalized treatment.

The analysis of gene expression data [2] is not so straightforward because the data dimensionality and the mode of obtaining differ significantly from standard data. The most important specificity is related to the availability of the high number of features compared with the number of instances. Typically, such a dataset consists of only one hundred instances described by thousands of features. This characteristic is because all these numerous gene expression levels are extracted in a single run of a device, and the real challenge is to gather a coherent group of cases (patients). Also, the functioning of genes and their interactions are not fully revealed, so the datasets contain all known genes. Most of the features can be redundant for a given task, whereas other features can be correlated.

Decision trees, as efficient feature-selecting algorithms, seem to be well suited for gene expression analysis. However, in the literature, there is not a lot of successful applications of existing classical inducers, and rather inferior predictive performance is reported [3, 4]. Instead, decision tree ensembles[1] are proposed [5, 6], and they can offer a competitive classification quality. The ensembles can be applied for biomarker identification [7], but interpreting multiple generated models is rather difficult [8], especially if the aim of the study is to better understand the underlying biological processes.

---

[1]Ensemble methods use multiple learning algorithms (so-called committees of classifiers) to obtain better predictive performance, and decision trees are very popular component classifiers.

© Springer Nature Switzerland AG 2019
M. Kretowski, *Evolutionary Decision Trees in Large-Scale Data Mining*,
Studies in Big Data 59, https://doi.org/10.1007/978-3-030-21851-5_7

It seems that one possible explanation of this rather disappointing performance of single tree approaches to gene expression data could be the underfitting problem [9]. This can be observed when the number of learning instances is low and with an increasing number of features the probability of finding an accidental feature, which perfectly separates instances, grows hazardously. As a result, a perfect reclassification is observed, but the real performance is obviously deteriorated. The detection or elimination of such spurious and/or noisy features is not easy, so more robust and reliable split constructions in non-terminal nodes are necessary. In this chapter, the concept of a multi-test split is presented, and the global, evolutionary approach is applied to a multi-test tree induction. This enables us to mitigate the under-fitting problem at the cost of a slight increase in the decision tree's complexity. Potentially, it can also reveal the hidden regularities in the form of gene clusters that constitute multi-tests.

7.1   Multi-test Splits

Multi-test splits are first introduced in the MTDT (Multi-Test Decision Tree) system [10], where a classical univariate test in a non-terminal node is replaced by a group of univariate tests working together. In other words, a multi-test split is based on a non-empty set of univariate tests, and the outcomes of a multi-test are decided using the majority voting of unit tests. The way in which the multi-tests work is similar to the ensemble's work but on a much more detailed level. It is assumed that the tests in a multi-test should behave in a similar way, meaning that they characterize a common phenomenon or process. It has been shown that polypeptides or proteins could be encoded by a group of functionally related genes [11]. This means that the identification of such groups and building multi-tests on them can bring not only better performance, but also new and curious patterns. The multi-tests should give more stable splits and allow for the elimination of the impact of incidental features.

All the tests in a multi-test need to have the same number of outcomes, which could be a restriction. However, in the case of gene expression data, where all the features are numerical, only standard inequality tests with two outcomes are utilized. The number of tests in a multi-test can be varied, but an odd number of unit tests is preferred because it eliminates the problem of a tie during the majority voting. The maximal number of unit tests should be imposed because too large multi-tests become rather unwelcome. Generally, the multi-test, which is formally a multivariate split, can be more difficult to interpret than a single test. But because it is composed of univariate tests, it could be perceived as an axis parallel split. In fact, for any multi-test, it is possible to create the corresponding decision subtree with only standard univariate tests. The multi-test should not only be treated as a compact representation of such a subtree because the possibility of identifying the groups of (functionally related) genes is also important.

As pointed out earlier, the tests in a multi-test are expected to be similar, meaning that they route instances in an analogous way. It is worth mentioning that in the

CART system [12], a mechanism for searching a surrogate test, has been proposed. It is applied when a (primary) test requires a feature value that is missing for an analyzed instance. In such a situation, a surrogate test on another available feature is applied. The surrogate test should best mimic the primary test in terms of directing the training instances to the corresponding branches. Hence, for every possible surrogate, the number of training instances, which are routed in the same way as the primary test, is calculated, and the surrogate with the highest score is selected. We decide to adopt this nomenclature in a multi-test definition. So among the unit tests, one test is selected and called a primary test. This test determines the purpose of the multi-test and is a reference point for the remaining tests, which are called surrogate tests. It should be noted that the tests in a multi-test work as a team, and during the majority voting, all the tests are equally important, so it is possible that a primary test is outvoted by its surrogates. Moreover, it does not mean that the best test (e.g., in terms of the gain ratio) will be the best primary test.

The similarity of tests can be quantified by the resemblance measure [9]. In a non-terminal node T_i with the corresponding test mt_i composed of a primary test pt_i and surrogate tests st_{ij}, the resemblance r_{ij} of j surrogate test to the primary test is defined as a fraction of the training instances that are directed in the same way as in two tests to all training instances in this node. When the tests with two outcomes are considered, it is possible to invert a surrogate test to improve the test resemblance r_{ij}. An example of such an operation is presented in Fig. 7.1. After inversion, the new resemblance is equal to $1 - r_{ij}$, which means that all surrogate tests with a resemblance lower than 0.5 should be inverted. Concerning the whole multi-test resemblance $Res(T_i)$, it can be simply calculated as the sum of the resemblance of all surrogate tests. Such a definition in contrast to, for example, the average resemblance, tries to take into account the size of a multi-test, so bigger and homogenous multi-tests can be promoted. For a multi-test reduced to just one test, the resemblance is equal to zero. So for an ideal multi-test composed of K tests, the resemblance is $K - 1$.

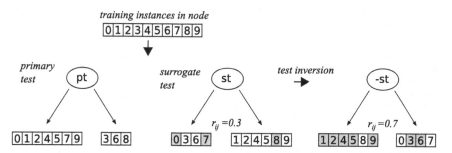

Fig. 7.1 An example of a surrogate test inversion to improve the test resemblance in a multi-test

7.2   Global Induction of Multi-test Trees

In the MTDT system, a multi-test tree is induced in a top-down fashion, and all multi-tests have the same size. Knowing the advantages of the global approach, it is clear that an evolutionary induction can bring important improvements. Searching for the best multi-test splits is embedded into the evolution algorithm, and it naturally leads to diverse sizes of multi-tests. The necessary extensions of the GDT framework are rather straightforward. The algorithm requires a slightly changed representation, the related modifications of genetic operators, and especially, an adaptation of the fitness function.

7.2.1   Representation and Initialisation

As can be expected, the only difference in a decision tree's representation is related to the replacement of simple tests by multi-tests in non-terminal nodes. Only inequality tests with two outcomes are considered as component tests, and the maximal number of tests in a multi-test during an initialization is fixed (default value: nine).

A high number of features (genes) is characteristic of the analyzed data. A lot of these features are superfluous or not related to the goals of the prediction task. Undoubtedly, it slows down the mining process, and various approaches can be proposed to alleviate the problem, like, for example, very popular initial feature selection [13]. We decide to incorporate a ranking of genes, which is used to set a preference for potentially more discriminative genes in numerous, random selections of the evolutionary algorithm. In this way, any features are arbitrarily excluded, but the search is somehow boosted. On the other hand, it is possible that even low-ranked genes can contribute and form interesting patterns, especially in lower parts of trees. For the ranking calculation, any algorithm that organizes genes according to some criterion revealing their discriminative power can be applied. Here, the most obvious choice is Relief-F [14], which is commonly applied in the feature selection of gene expression data [15]. The ranks for all the features are precalculated on the whole dataset before the actual algorithm is run. These ranks are scaled to the [0, 1] range, where zero corresponds to the highest ranked gene. The assigned values can be interpreted as a kind of cost, which means that less-costly features will be chosen more often.

An initial population of decision trees is generated as usual by applying a top-down method for randomly selected training sub-samples. Initial multi-test splits are created with the following simple heuristic: In the first stage, a primary test is searched in two steps: a gene is randomly selected from the ranked list of genes, and then, for this feature, all candidate thresholds are processed. A list of thresholds is sorted according to their gain ratio, and the second draw is performed to complete the primary test. In both these draws, the (exponential) ranking selection is used. This method clearly prefers the best genes and thresholds, but also, low-ranking choices are possible.

The second stage is devoted to the generation of the corresponding surrogate tests. First, their (even) number is randomly chosen, and each test is created based on a randomly selected (but different) feature. However, this time, the choice of thresholds needs to be modified because the surrogate tests should try to closely mimic the primary test. Hence, to rank the thresholds, the corresponding surrogate test resemblance is used. As a result, the tests similar to the primary test are more likely to be selected.

7.2.2 Genetic Operators

The introduction of multi-test splits implies the use of necessary mutation operator extensions in the modification of the multi-tests themselves. Other minor adjustments of the remaining genetic operator variants will not be discussed. In fact, the existing multi-test in an internal node can be replaced by a new multi-test (generated in the same way as during the initialization) or modified. There are different variants of the multi-test modifications, but they all accept the leading role of the primary test, and it is the primary test that triggers the modifications of surrogate tests:

- The current primary test can be modified by its threshold shifting as in the standard univariate test. However, this entails the need for adjusting all the thresholds in the surrogate tests. It requires some computation, but allows for improving the resemblance and should be perceived as a memetic extension.
- The test roles in a multi-test can be randomly exchanged. This means that one of the surrogate tests becomes a new primary test, and the former primary one becomes a surrogate. This may require the surrogate tests to adjust their thresholds to better fit to the primary test. It is also a memetic-like variant.
- The number of surrogate tests in a multi-test can be increased by two on the condition that the maximal number of tests is not reached. Two additional tests are created in the same way as in the initialization, and they have to introduce novel features into the multi-test.
- The number of surrogate tests can be decreased by two. The eliminated tests are randomly selected based on a (small) ranking using their resemblance.
- A randomly selected surrogate test can be replaced by a new test. The same ranking as in the previous variant is used.

The presented set of multi-test mutation variants tries to limit the search space of the multi-tests by focusing on finding the primary tests. The surrogate tests are largely dependent on their primary tests, so only the number of surrogate tests is actually fully evolved. When searching for the surrogate tests, the local search components are employed extensively.

7.2.3 Fitness Function

For typical datasets, data overfitting is the most severe problem that learning algorithms have to face. The algorithms tend to create very complex and/or detailed predictive structures that work pretty well for known instances, but they lose their precision for new data. A lot of effort is spent on making the predictors more general, which usually improves the actual performance. In the case of evolutionary decision tree induction, the complexity term in the fitness function is introduced to prevent overspecialization. However, for gene expression data, this type of approach is not appropriate, because here, the potential danger is associated with underfitting. The problem of not having a proper generalization needs to be attacked from the other side. With the small number of instances and large number features, the splits can be seemingly good but in fact accidental. This can lead to very simple trees that predict poorly and do not offer any insight into the analyzed data.

The introduction of a multi-test in a tree's representation alone does not solve the problem. The fitness function needs to be revisited and adjusted to account for the specific requirements. First, one should be interested in multi-tests that are possibly coherent and relatively big. Hopefully, the multi-test resemblance reflects these objectives well. It also seems that highly ranked genes should be promoted because this could accelerate the evolution convergence and reduce the risk of including noisy features, especially in the lower parts of trees, where the number of available training instances is further reduced. Hence, a combination of these two measures will replace the original complexity term. Obviously, the low reclassification error should remain among the objectives, but this can be expected; for most of the datasets, the error will be minimal. Finally, the modified fitness function can be formulated as a weighted sum:

$$Fitness(T) = Err(T) - \alpha * Res(T) + \beta * Rank(T), \qquad (7.1)$$

where $Res(T)$ is the overall resemblance of the multi-tests, and $Rank(T)$ is the overall cost of the features defining the multi-tests. Parameters α and β are the relative importance of the resemblance and ranking components, and their default values are 1.0 and 0.5, respectively.

The last element that should be specified is how to combine the partial characteristics describing each multi-test into the overall measures of the decision tree. Let $\xi(T_i)$ be a fraction of the training dataset in the T_i node (the number of instances in the T_i node divided by the whole training dataset size) and $IN(T)$ a set of non-terminal nodes in the T tree. For the root node, $\xi(T_i)$ is equal to one, and for all remaining nodes is lower. The overall resemblance $Res(T)$ can be defined as:

$$Res(T) = \sum_{T_i \in IN(T)} \xi(T_i) * Res(T_i). \qquad (7.2)$$

The rationale for such a definition is that it takes into account the hierarchical structure of a tree, and the relative importance of a multi-test is associated with the size of the training subset on which the test has been built. In the case of the ranking component, a similar idea is followed, but because the meaning of this measure is different, this time, the relative importance of the multi-test features in the lower parts of a tree is increased.

$$Rank(T) = \sum_{T_i \in IN(T)} \frac{1}{\xi(T_i)} * Rank(T_i). \tag{7.3}$$

This could help prevent the classical overfitting because it will possibly limit the multi-test sizes built on a small number of instances.

7.3 Experimental Results

The adequacy of the presented extension of global decision tree induction for mining the gene expression data needs to be verified experimentally. The performance of the GDT-based multi-test tree evolution is compared with the results obtained by the following three systems:

- CART [12]—a representative of the classical top-down decision tree systems. The SimpleCART implementation from the WEKA toolkit [16] is used. The other algorithms available in this toolkit (e.g., J48 and REP Tree) are also launched, but the results are slightly worse.
- MTDT [10]—the top-down inducer in which multi-test trees are introduced; the (maximal) number of component tests in a multi-test is fixed.
- HEAD-DT [4]—the hyper-heuristic evolutionary system that searches for the best combination of the various components of classical top-down algorithms.

The analyzed gene expression datasets deal with various types of cancer and are first described in [17]. In [4], these thirty-five datasets are divided into two groups: fifteen datasets are used for tuning, and the rest for testing the algorithms. Here, such a setup is followed, but only the results obtained for the largest nine datasets (more than 1000 features and at least fifty instances) from the testing part are presented. In Table 7.1, the analyzed datasets are characterized.

A ten-fold cross-validation is repeated ten times, and the obtained results are averaged. As in [4], the HEAD-DT system is tuned on a separate group of datasets, while the other algorithms are tuned on the same data to enable a fair comparison. According to the recommendations from Auto-WEKA 2.0 [18], a large number of parameter values is tested, and an unpruned version demonstrates the highest average accuracy on the tuning package, but the gain compared with the baseline settings is marginal. Concerning the MTDT system, various sizes of multi-tests are tested, and the best performance is observed for five component tests. Regarding the modified GDT, only the influence of two coefficients (α and β) from the fitness function is studied. For both parameters, the range [0, 2] is tested. In Fig. 7.2, the impact

Table 7.1 The characteristics of gene expression datasets

Dataset	Instances	Features	Classes
Alizadeh-v2	62	2092	3
Alizadeh-v3	62	2092	4
Armstrong-v1	72	1080	2
Bhattacharjee	203	1542	5
Golub-v1	72	1867	2
Lapointe-v1	69	1624	3
Lapointe-v2	110	2495	4
Nutt-v1	50	1376	4
Tomlins-v1	104	2314	5

Fig. 7.2 The impact of α and β on the average accuracy of global multi-test tree induction estimated on the tuning package

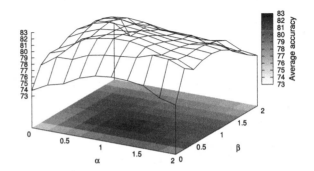

of these two parameters on the average classification accuracy observed on all the tuning datasets is presented. The performance is rather stable even when using a wide range of parameter values. Here, the performance decreases only for extreme parameter values (especially for a small β).

In Table 7.2, the results obtained by four inducers are presented. It can be observed that the modified GDT system obtains the best accuracy for seven out of nine the datasets (being the sole leader for five datasets). Interestingly, in the case of the *Lapointe-v2* dataset, the best accuracy is attributed to MTDT, and only for one dataset (namely *Alizadeh-v3*) does HEAD-DT beat its counterparts. This suggests that the global induction combined with multi-test splits offers some advantages over univariate top-down inducers when it comes to gene expression data.

It may also be interesting to confront the complexity of the induced classifiers. In Table 7.3, the average number of nodes is presented, except for HEAD-DT, because the tree size is not provided in [4]. For evolutionary-induced multi-test trees, the average number of component tests is also given, whereas for the MTDT, the size of the multi-tests is constant.

Here, the number of univariate tests is visibly increased in multi-test trees compared with standard trees, and this is a rather expected effect. On the other hand, the tree size expressed as the number of nodes is systematically lower for the multi-test

Table 7.2 Accuracy obtained for gene expression datasets

Dataset	GDT	HEAD-DT	MTDT	CART
Alizadeh-v2	0.96	0.88	0.94	0.89
Alizadeh-v3	0.71	0.74	0.71	0.71
Armstrong-v1	0.95	0.90	0.93	0.89
Bhattacharjee	0.90	0.90	0.90	0.90
Golub-v1	0.95	0.88	0.93	0.85
Lapointe-v1	0.73	0.66	0.72	0.70
Lapointe-v2	0.69	0.62	0.72	0.63
Nutt-v1	0.54	0.53	0.54	0.54
Tomlins-v1	0.66	0.59	0.64	0.60

Table 7.3 The average tree size obtained for gene expression datasets. In brackets, the average number of tests in the multi-tests is given

Dataset	GDT	MTDT	CART
Alizadeh-v2	3.2(5.9)	3.0	5.0
Alizadeh-v3	6.3(4.9)	5.6	9.0
Armstrong-v1	2.1(7.2)	2.7	3.8
Bhattacharjee	7.4(6.9)	8.8	12.2
Golub-v1	2.0(3.1)	3.2	4.6
Lapointe-v1	5.9(3.5)	7.0	9.6
Lapointe-v2	10.7(3.4)	13.4	16.9
Nutt-v1	6.8(3.4)	6.9	10.5
Tomlins-v1	11.5(3.8)	12.2	18.8

trees. The average number of component tests never drops below three, which shows that evolutionary induction finds a lot of coherent gene clusters.

Let us concentrate on the first dataset to better understand the dynamics of the evolutionary induction and the roles of resemblance and feature ranking in it. This will also provide some biological insights. The *Alizadeh-v2* dataset [19] deals with three of the most prevalent adult lymphoid malignancies: diffuse large B-cell lymphoma (DLBCL), follicular lymphoma (FL), and chronic lymphocytic leukemia. In Fig. 7.3, the course of evolution is presented. The various characteristics of the best individual are tracked through 6000 iterations. The results are averaged over fifty runs, so the curves are smoother. Two variants of the global induction are studied -with feature ranking and without feature ranking (all features are equally probable in every draw).

Figure 7.3a illustrates the transition of accuracy measured with the testing parts during the evolution. The reclassification accuracy is not presented because it immediately reaches one hundred percent. Here, the ranking-based version offers better accuracy, and a gap grows with the subsequent iterations. Interestingly, it seems that the proposed initialization is very good for this dataset because the initial accuracy

Fig. 7.3 The dynamics of
evolutionary multi-test tree
induction. The comparison
of the evolution course with
and without using feature
ranking: **a** test accuracy and
average resemblance of
component tests, **b** tree size
and average multi-test size,
and **c** fraction of top- and
low-ranked features in the
tests

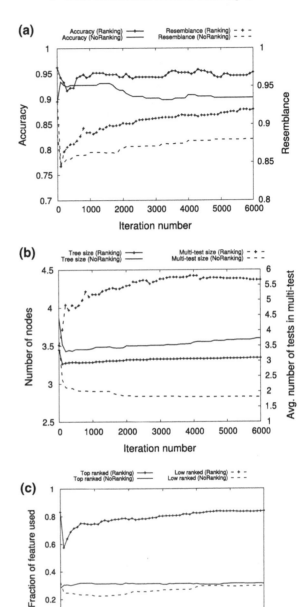

does not change much. Moreover, the average resemblance of the component tests in the multi-tests can be observed. As expected, the resemblance constantly increases, and the feature ranking clearly helps to find coherent tests.

Figure 7.3b concentrates on the sizes of the trees and multi-tests. It can be noticed that the proper tree size is found early and remains stable. The use of feature ranking results in smaller trees in terms of the number of nodes. The difference between the two considered versions is much more visible for the average multi-test size. During the evolution, the separation between them grows, and finally, the size of the multi-tests is three times larger with feature ranking enabled.

The last panel in Fig. 7.3 confirms that feature ranking really works. The important difference in the top-ranked and low-ranked feature usage is highly correlated with the algorithm version. The use of the ranking actually promotes top-ranked features and reduces low-ranked ones.

In Fig. 7.4, an example of an evolutionary-induced multi-test decision tree for the *Alizadeh-v2* dataset is presented. The tree consists of just one multi-test composed of seven univariate tests and two additional simple tests. For the both training and testing data, a perfect accuracy is obtained.

Based on GenBank [20] information, among the nine genes appearing in the tests, seven are known as directly related to lymphoid malignancies. For example, the PCDH9 gene (#51439) is downregulated in non-nodal mantle cell lymphoma and glioblastoma as a result of gene copy number alterations [21]. Interestingly, two tests (#701796 and #714453) from the root node multi-test are related to a novel driver alternation IL4R [22] for the subtype of DLBCL called primary mediastinal large B-cell lymphoma.

A more detailed analysis (comprising all thirty-five datasets, more inducers, statistical analysis of the results, and a discussion of the biological interpretations of some case studies) is presented in [9].

Fig. 7.4 An example of an evolutionary-induced multi-test decision tree for the *Alizadeh-v2* dataset

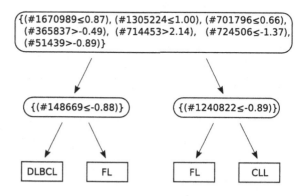

References

1. Pevsner J (2015) Bioinformatics and functional genomics. John Wiley, Hoboken
2. Bellazzi R, Zupan B (2007) J Biomed Inform 40(6):787–802
3. Grzes M, Kretowski M (2007) Biocybern Biomed Eng 27(3):29–42
4. Barros R, Basgalupp M, Freitas A, Carvalho A (2014) IEEE Trans Evol Comput 18(6):873–892
5. Chen X, Ishwaran H (2012) Genomics 99(6):323–329
6. Lu H, Yang L, Yan K, Xue Y, Gao Z (2017) Neurocomputing 228:270–276
7. Lazzarini N, Bacardit J (2017) BMC Bioinform 18:322
8. Nag K, Pal N (2016) IEEE Trans Cybern 46(2):499–510
9. Czajkowski M, Kretowski M (2020) Expert Syst Appl (in review)
10. Czajkowski M, Kretowski M (2014) Inf Sci 288:153–173
11. Yi G, Sze S, Thon M (2007) Bioinformatics 23(9):1053–1060
12. Breiman L, Friedman J, Olshen R, Stone C (1984) Classification and regression trees. Wadsworth and Brooks, Monterey
13. Guyon I, Elisseeff A (2003) J Mach Learn Res 3:1157–1182
14. Robnik-Sikonja M, Kononenko I (2003) Mach Learn 53(1–2):23–69
15. Hira Z, Gillies D (2015) Adv Bioinform 2015:198363
16. Frank E, Hall M, Witten I (2016) The WEKA workbench. Online appendix for "Data mining: practical machine learning tools and techniques", 4th edn. Morgan Kaufmann, Burlington
17. de Souto M, Costa I, de Araujo D, Ludermir T, Schliep A (2008) BMC Bioinform 9:497
18. Kotthoff L, Thornton C, Hoos H, Hutter F, Leyton-Brown K (2017) J Mach Learn Res 18(1):826–830
19. Alizadeh A et al (2000) Nature 403:503–511
20. Benson D, Cavanaugh M, Clark K, Karsch-Mizrachi I, Ostell J, Pruitt K, Sayers E (2017) NuclC Acids Res 45(D1):37–42
21. Wang C et al (2012) J Clin Neurosci 19(4):541–545
22. Vigano E et al (2018) Blood 131(18):2036–2046

Part IV
Large-Scale Mining

Chapter 8
Parallel Computations for Evolutionary Induction

Top-down decision tree inducers are very fast, and even if the obtained decision structures are just good and not really close to the possible optimal solutions, it is often enough for practitioners who are interested in solving specific problems. The implementations of the most popular greedy algorithms are available in every data mining commercial system, and they can be very easily applied without any profound awareness of the parameter settings or running details. Moreover, knowledge of the existence of alternative induction methods is still limited to a narrow group of researchers who are working on this topic.

Evolutionary-based data mining systems can produce more concise and accurate predictors for many datasets, but they require many more computational resources. The problem becomes even more obvious when large-scale data applications are considered. On the other hand, in recent years, there has been significant progress both in specialized computing hardware solutions and software platforms facilitating their usage. This process should be perceived as a great opportunity for computationally intensive approaches.

As expected, a real breakthrough in the popularity of evolutionary induction is not possible if more efficient implementations do not appear and become accessible outside of data scientists. Hopefully, it is well-known that population-based heuristics are naturally prone to parallelization [1] because they process many candidate solutions at one time. Especially, a lot of effort has been put into the distribution/parallelization of an evolutionary search [2, 3]. A lot of interesting results have been published on the structuralization of a population within island models or cellular approaches, where improvements in algorithm efficacy are the most important goals.[1] In this book, however, I concentrate only on so-called global parallelization with the aim of as much acceleration of the evolutionary decision tree induction as possible. It should enable the actual processing of large-scale data on affordable computing platforms,

---

[1] Some results on the distributed global induction of decision trees within a multi-population island model can be found in [4].

© Springer Nature Switzerland AG 2019
M. Kretowski, *Evolutionary Decision Trees in Large-Scale Data Mining*,
Studies in Big Data 59, https://doi.org/10.1007/978-3-030-21851-5_8

like commodity clusters or stronger GPU-equipped stations. It seems that GPU-based solutions are especially well suited for speeding up an evolutionary search [5, 6]. They can ensure the success of the global approach much like they have done with deep learning.

In the case of the evolutionary induction of decision trees, there are two natural levels of parallelization. The first one is simply linked to any population-based algorithm; in many of these situations, individuals can be treated as independent entities and processed in parallel. Typically, the most time-consuming operation is the evaluation of the fitness of all the candidate solutions. This process can be easily parallelized with the well-known master-slave approach by distributing the individuals evenly among the available computing processors (slaves), whereas the master processor executes the remaining operations of the evolution. The upper limit of such a speedup is bounded with the number of individuals in a population (usually between fifty and one hundred). The second source of parallelization could be derived from the dataset. In most induction algorithms, the fitness function is somehow related to the predictor accuracy estimated on the learning (or validation) datasets. This gives an opportunity to apply a classical reduction pattern, where the performance is first assessed on subsets and then aggregated. When this source of parallelization is exploited, the upper bound of the speedup is obviously much higher, especially for large-scale data and massive computation equipment.

In this chapter, three different approaches for speeding up the evolutionary induction of decision trees are presented. First, hierarchical MPI+OpenMP parallelization is applied as a typical solution for computing clusters. Next, the GPU-based approaches for single and multiple cards are described. The last solution uses Apache Spark. For all the techniques discussed, the results of the experiments using large-scale data are provided.

8.1 Parallel Implementation with MPI+OpenMP

For many data mining tasks, the time complexity of the applied tools is an equally important factor as their prognostic performance. This is especially visible for large-scale data when the learning of the predictors is inevitably long(-er). In the case of the sequential implementation of an evolutionary induction, the dynamic program analysis clearly shows that the permanently available information about the location of the learning objects in the tree nodes is crucial for time efficiency. As a result, with every individual representing a decision tree, an additional table,[2] with the indexes of the objects located in each node, is maintained. Based on the location info, the genetic operators can be applied in a targeted way, and the fitness can be evaluated instantly. After every effective mutation or crossover, the locations of the learning objects only in the affected parts of the tree(s) need to be restored. Hence, the actual fitness calculation is much faster [7, 8] and can be embedded into the post-

---

[2]The size of the table is equal to the number of objects in the learning dataset.

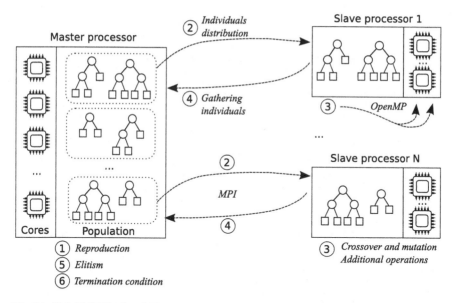

Fig. 8.1 Hybrid (MPI+OpenMP) parallelization of the global decision tree induction based on the master-slave model

genetic operators' processing. This mechanism increases the memory complexity of the algorithm but significantly reduces its computational complexity.

Even with very reasonable improvements, the sequential version of evolutionary induction is not the fastest, especially if compared with top-down greedy inducers. It is natural that the possibilities for speeding up the computation are investigated. In such a situation, the use of computing clusters built using many nodes with multi-core processors is often considered as a possible, first-choice solution but with the condition that the user has access to such equipment. Usually, for communication between the nodes of such a system, the message-passing strategy is applied, whereas inside the node, the shared memory is the best communication channel.

The parallel solution described here is based on a typical master-slave approach, and it combines a Message-Passing Interface with OpenMP. The first attempt to parallelize the GDT solution is proposed in [9]. It is the pure MPI-based version applied only to classification problems. The extended hybrid version is better suited for current hardware and can be applied to classification [10] and regression [11] problems. Figure 8.1 illustrates the presented parallel approach. In the first step, the master node spreads the individuals from the population over the slave nodes using the MPI. Next, in each slave node, the calculations are spread over the available cores using OpenMP, and the algorithm blocks are run in parallel.

8.1.1 Distribution of Work into Processes

In each evolutionary loop, the master evenly distributes the individuals between the nodes (slaves). To avoid wasting resources, a chunk of the population is left on the master, which also works as a slave. Migration of the individuals between nodes is performed within the framework of the message-passing interface and requires the following:

- Packing the tree structures into a flat message.
- Transferring the message between nodes (sending/receiving).
- Unpacking the message into the corresponding tree.

The packed tree structure contains information about its size, the tests in the internal nodes, and additional nodes statistics (e.g., the number of learning vectors), which speeds up the reconstruction during the message unpacking in the target node. To minimize the message size, the information about the learning vectors associated with the tree nodes is not included in the message.

The certain parts of EA such as reproduction (with elitism) and termination condition verification are executed on the master node. However, to perform the selection, the fitness value of the distributed individuals must be known. To avoid unnecessary unpacking-packing operations on the master (for trees that will not be selected in the next generation), the fitness value of the migrated individual is also transferred. Additionally, a certain number of individuals from the given slave node may survive (or be replicated), and in the next iteration, they could be scheduled to be sent back to that node. This gives another possibility to eliminate unproductive calculations. However, it raises the risk that the individuals in a particular node may not change enough (or will be very similar). To keep the original sequential algorithm and avoid some kind of island model, we send the cloned trees to random slave nodes, hence keeping the sub-population diversified at each slave and avoiding crossing with identical or very similar individuals.

In the previously presented research [12] on parallel GDT implementation, the redistribution of learning vectors was performed after unpacking each individual on the target node. The whole tree is reconstructed before starting the mutation and crossover operations. Then, after a successful application of a genetic operator, the redistribution of the learning vectors of an affected node (and eventual sub-nodes) is performed.

This process of associating each instance with the appropriate leaf is very time-consuming, especially on large datasets. To limit the redistribution of the data, we propose reconstructing only those nodes that will be affected. In addition, there is only the need to find the learning instances that fall into the affected node because the redistribution of the eventual sub-nodes is not necessary. If a genetic operator will be successful, the learning instances in the sub-tree will be relocated anyway. In this way, instead of reallocating all the learning objects in the whole tree, we only set a part of the data into the node (without its eventual sub-nodes) that is selected for a mutation or a crossover. If the root tree node is to be affected by a genetic operator, the preceding processing is highly reduced. Only, the whole dataset is associated with the root node. The GDT assumption that internal nodes in the lower parts of the tree are mutated with a higher probability also enhances the possible speedup of the proposed implementation. It can be expected that the lower parts of the tree will hold fewer learning objects that need to be relocated.

8.1.2 Distribution of Process Work to Threads

The shared memory approach is applied in every slave node (including the master, which works also as a slave). It is assumed that all the cores within a node operate independently but share the same memory resources. Access and modification of the same memory space by one core is visible to all other cores; therefore, no explicit data communication between the cores is required. However, additional synchronization during write/read operations is needed to ensure there is appropriate access to shared memory.

In Fig. 8.1, each slave node spreads the calculations further. The calculations in the chosen algorithm blocks concerning different individuals are spread over the cores. In this way, all types of genetic operations together with the redistribution of learning objects can be performed in parallel. In the case of a mutation, each core processes a single individual at a time, whereas during crossover, the pairs of affected individuals are processed in parallel. Parallelization with a shared memory approach is also performed on the master node for the distribution of the population to other nodes and for gathering the population from them. In addition, all the trees that were pruned into leaves after the application of genetic operators, are extended into sub-trees in parallel by the cores at each slave node.

8.1.3 Experimental Results

Because the parallel implementation does not change the actual induction algorithm, the presented results are focused on the obtained speedup. The only differences are because of the draws (of pseudo-random numbers) that need to be performed in a decentralized way, but the evolutionary induction is highly stochastic, and it does not really influence the obtained predictions.

Experiments are performed on a cluster of sixteen SMP[3] servers (nodes) running Ubuntu 12 and connected by an Infiniband network (20 Gb/s). Each server is equipped with 16 GB RAM, two processors Xeon X5355 2.66 GHz with the total number of cores equalling eight. As a software tools the Intel C++ compiler (version 15.1), MVAPICH version 2.2, and OpenMP version 3.0 are used.

---

[3]Symmetric multiprocessing (SMP)—multiprocessor machine with a shared main memory that is controlled by a single operating system instance that treats all the processors (cores) equally.

Table 8.1 Mean speedup for different numbers of cores. The first two datasets are classification problems, whereas the remaining are regression problems

Dataset	Instances	Features	Speedup for number of cores		
			4	16	64
Chess3x3	100 000	2	2.75	8.06	15.34
Zebra	100 000	10	2.61	5.60	9.93
Fried	40 768	10	2.81	8.54	23.08
Elnino	178 080	9	2.82	7.82	15.80

Table 8.1 presents the obtained mean speedup for a few exemplar (both classification and regression) artificial and real datasets from the UCI Machine Learning Repository [13]. For classification, the mixed version of the GDT system is applied, while for the regression problems, two univariate variants are considered: a simplest regression tree and model tree with multivariate linear models in leaves. The maximal number of OpenMP threads per node is restricted to eight (or to the number of overall cores for the smallest configurations) because in the available hardware, there are eight (2×4) cores in each node. The parallel implementations are run with the same (default settings) parameters as the corresponding sequential versions. In Table 8.1, only the results for the best combination of nodes and cores (MPI processes and OpenMP threads) are presented.

Here, for all these datasets and decision tree variants, the obtained speedup for four cores is at least acceptable (2.5–3) when taking into account the unavoidable overhead and that certain algorithm parts remain sequential. Obviously, it is a shared memory, which helps in obtaining this result. For more cores, the efficiency degrades as expected when more communication between the nodes through message passing is necessary. For sixty-four cores, the differences between the best and worst speedup are already remarkable (ten to twenty-three), but even ten can be quite useful in practical applications because it reduces the induction time from hours to minutes.

In Fig. 8.2, the scaling properties of the hybrid parallel implementations are presented. For every number of cores, the best combination of the number of MPI processes and OpenMP threads is used for a given dataset (e.g., for sixteen cores and 10 000 objects variant of *chess3x3* dataset, the best configuration is four nodes with one MPI process per node and four OpenMP threads inside each node, whereas for 1 000 000 variant, it is eight nodes with one MPI process and two OpenMP threads, respectively). It can be observed that the obtained speedups, regardless of the datasets' sizes, are similar for every type of the tree. For the model tree induction, the observed speedup is better, but this can be easily explained. In a model tree induction, a very important part of the processing time (even up to sixty percent of the evolutionary iteration time) is spent on building the linear models in the leaves and calculating the predictions. This part of the algorithm parallelizes very well, so the sequential fraction is reduced compared with a simpler regression tree induction.

Fig. 8.2 Speedup evaluation for parallel implementations of classification (*cla.*), regression (*reg.*), and model (*mod.*) tree induction. The variants of *Chess3x3* (classification) and *Armchair* (regression) datasets are analyzed with a varying size from 10 000 to 1 000 000 objects

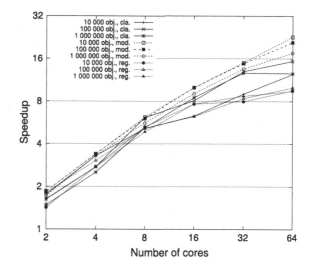

Additionally, the overhead because of the MPI transfer is considered, and as a result, according to Amdahl's law, the efficiency for a higher number of cores decreases.

8.2   GPU-Based Acceleration

The GPU is especially well suited for performing thousands of the same, simple calculations in parallel,[4] preferably on independent elements. Generally, in evolutionary decision tree induction, the most time-consuming operation is calculating the individuals' fitness because this requires each time processing the whole training dataset. Moreover, for model trees, fitting the models in leaves adds substantial computational complexity. To estimate a prediction error, every object from the training dataset is passed through the decision tree, starting from the root node until reaching one of the leaves. In each step, a test is performed, and its outcome redirects the object to the next node. If the tree is full, the number of tests is the same and equals to the tree depth. Otherwise, the number of tests can be different for various objects, but typically, such a difference is not large. From a parallelization point of view, it is especially useful that every object can be processed apart from one another. This observation directly leads to the idea that the distinct parts of the training dataset can be processed at the same moment. In Fig. 8.3, the applied two-level decomposition strategy of the dataset is presented. At first, the whole dataset is divided into smaller, equal-sized data subsets that are processed by different GPU blocks. Next, in each block, the objects from the data subset are spread further over the threads.

---

[4] Avoiding the conditional statements allows for a reduction in the warp divergence.

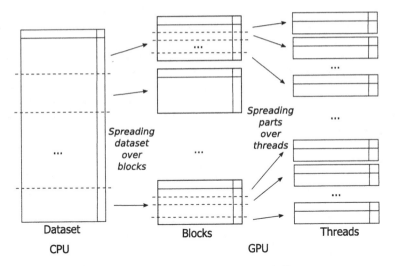

Fig. 8.3 Datasets' decomposition into blocks and threads on the GPU

Because the transfer from the CPU to GPU is slow and the training dataset is used constantly during the calculations, the dataset is preloaded onto an allocated space in the global GPU memory, and it can be accessed by all the blocks through out the entire evolution.

Based on the presented decomposition strategy, the two-step algorithmic pattern of confronting the full dataset with a particular tree can be distinguished (Fig. 8.4). Each GPU block receives a copy of the processed individual with additional accumulators (that are specific to the decision tree type), and such a structure is loaded onto the shared memory that is visible by all the threads within the block. In this way, every thread works on a separate data chunk, but the threads from one block can work in parallel and update the same copy. When needed, the threads in one block are synchronized using atomic operations.

In the first step, all the objects from a data chunk corresponding to one thread are sequentially processed by traversing the tree from a root node until reaching one of the leaves. Each time an object reaches the final node, an appropriate accumulator (or accumulators) is updated. When all the blocks terminate processing, the second step can be performed. All copies of the tree with accumulators from the blocks are reduced into just one. The aforementioned pattern is typically implemented by two consecutive GPU kernels [14], and it can be applied to calculate the various tree characteristics depending on the tree type. In the case of classification trees, the reclassification accuracy is needed, so here, we are interested in revealing the predicted classes (decisions) that are associated with each leaf. For this, the relative frequencies of the classes of the training objects in leaves need to be known because the majority rule is applied in every leaf. Based on the frequency of the predicted class and the frequencies of other classes in each leaf, the reclassification accuracy of the tree can be estimated.

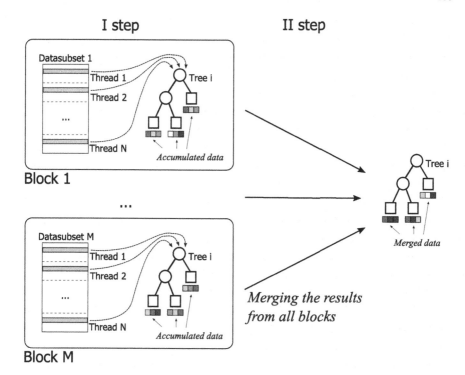

Fig. 8.4 Typical algorithmic template that can be applied to calculate the various tree characteristics based on the dataset

In the case of the simplest regression trees [15], for estimating the root mean squared error on the training dataset, this algorithmic pattern must be applied twice. First, it is done for estimating the predictions (averages of the target variable) in leaves. And second, it is done for calculating the prediction error on the training dataset. In the first case, instead of class counters, the accumulators are used to sum the target feature values and the number of training objects in every leaf. In the second case, the squared residuals are accumulated.[5]

Regarding model trees, the situation is much more complicated. In the GDT system, it is assumed that the (multivariate, linear) models in leaves are not evolved, but they are directly dependent on the selected features (or just one feature) and the training objects that reached the corresponding terminal nodes. In the sequential version, the locations of all the objects from the training dataset are memorized and maintained for each individual, so the models are recalculated only when a set of objects in a leaf is changed. In the GPU-based version, the locations of the training instances are not kept on the CPU side, which significantly reduces both the

---

[5]It may be possible to simplify the second stage by temporal memorization of the training instance locations in a tree after the first stage. Indeed, this concept is implemented for the GPU-based model tree induction.

computational and memory complexities of the code, which here is not parallelized. For model trees, this results in the necessity of providing which models are not valid after genetic operator applications and hence should be refitted. It is also possible just to recalculate all the models when a tree is evaluated. This second solution is algorithmically simpler, but obviously, it is more computationally demanding. Fortunately, fitting a linear model in a leaf can be done efficiently with the use of the optimized parallel procedures from the available libraries. However, each time, the training instances located in a leaf need to be known as an input.

Hence, the evaluation of a model tree in the GPU is organized in the following four stages using a few kernel-based procedures:

1. The whole dataset is passed through a tree. The same mechanism is applied as in the aforementioned simpler tree processing method, but instead of using accumulators, the locations of the training objects are stored in the global memory (as pairs of an object number and a leaf number).
2. The obtained data are reorganized in the GPU memory to be ready for the necessary fitting of the models in leaves with specific library procedures. This processing includes arranging key-value (leaf number–object number) pairs with a radix sort algorithm.[6] Furthermore, it encompasses allocating auxiliary matrices, copying, and transposing instances.
3. Calculating the necessary linear models in each leaf is performed leaf by leaf by using the linear algebra GPU-accelerated procedures (QR factorization and back-solving linear equations).[7]
4. When the models in leaves are ready (recalculated or sent back from the CPU) and because the locations of the training instances were stored after the first stage, the prediction error can be calculated in parallel in blocks, and then, merged based on the previously applied two-step mechanism. The corresponding model is launched on each object from the chunk, and the squared residuals are accumulated. Finally, the last kernel is used to reduce the errors from all the copies of the tree in the blocks.

Here, the GPU-based evaluation of the model trees is much more memory intensive because the locations of all the training instances are preserved between the stages, and many additional data structures need to be allocated to use optimized library procedures. As a result, the maximal dataset size, which can be analyzed with model trees, is reduced when compared with classification trees.

Unfortunately, the elimination of maintaining the training data locations on the CPU side poses a certain problem because for some variants of genetic operators dipoles are needed. For example, when a mutation is applied to a leaf and a new internal node is created, an effective test needs to be constructed. A suitable dipole (e.g., mixed for classification trees or long for regression trees) can be used to guide the

---

[6] A radix sort algorithm is considered to be one of the fastest sorting algorithms on the GPU [16]. The state-of-the-art GPU-based implementation of it is available in the CUB library [17].

[7] For linear algebra computations in the CUDA environment, highly optimized cuBLAS [18] and cuSOLVER [19] libraries are applied.

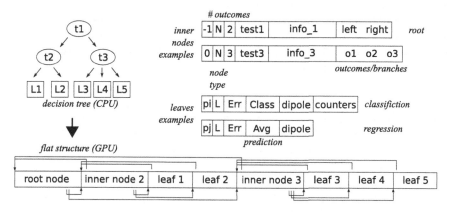

Fig. 8.5 Converting a decision tree to a flat representation composed of segments corresponding to the nodes used on the GPU. The examples of segments describe classification and (simplest) regression leaves and univariate non-terminal nodes

procedure. These dipoles should be randomly constructed from the training objects residing in a node that is chosen by an operator. In the GPU-based version, a bunch of potentially useful dipoles is always randomly chosen when the training dataset is passed through the tree when using the above-described kernel functions. In the first kernel, additionally, the objects constituting dipoles in each leaf are randomly chosen in each block. Next, when the copies of the tree are merged, the dipoles are randomly and simultaneously selected from them. Such a solution ensures that the computational overhead is minimal. Finally, when the tree is sent back to the CPU, the objects from the dipoles are also transferred, and they can be used in the next iteration.

It is worth to recalling at this point that the programming capabilities and memory organization are different between the CPU and GPU. As a result, an individual representation needs to be adjusted for the GPU. In Fig. 8.5, an example of a flat representation (one-dimensional array) is sketched. For each node, a fixed-size segment of the array with a structure depending on the node type (internal or terminal node) is associated with the references to the positions of parent and children nodes. Details of the segment's structures depend on the tree type. Any decision tree is converted into a flat representation before being transferred to the GPU, where this representation is used during computation. When the evaluated individual is sent back from the GPU, it is decoded, and the typical object-oriented representation of the decision tree is further processed.

From an implementation perspective, developing an efficient GPU-based variant of any algorithm is undoubtedly not an easy task because it requires not only high programming skills, but also up-to-date and profound knowledge concerning both the available hardware and software resources. On the other hand, the resulting payoff can be spectacular. Nowadays, very strong GPU cards can be acquired at reasonable prices and installed on standard personal computers, creating very competitive computing stations. It is especially evident if compared with the costs and capabilities of computing clusters. Of course, not all algorithms or applications are equally fitted to the GPU-based ecosystem, but in the case of evolutionary data mining, the observed price-to-effect ratio is excellent.

8.2.1 Speeding up the GPU Version

The speedup obtained with the GPU version is substantial, but if the developed algorithm is analyzed in detail, it becomes evident that many calculations are repeated. And this means that room for improvements still exists. There are at least two types of repeated calculations that can be eliminated. The first type exploits the similarity between the trees in two subsequent iterations. After a typical mutation, the resulting tree differs from the original tree only in the affected subtree. This observation enables limiting the propagation of an important part of instances in the evaluated tree, which is necessary to estimate the individual fitness [4]. In Fig. 8.6, a partial propagation of training instances is presented. Depending on the position of the modified node, the affected part of the tree can be very distinct. If the root node is mutated, the whole dataset needs to be propagated through the whole tree, but if the modified node is lower in the tree, the tree part that is not affected grows. In the most optimistic case, when an internal node is pruned into a leaf, only the predictions in this leaf need to be recalculated, and the statistics on the path to the root node are updated. Regardless of the position of the affected node, the test associated with the root has to be checked for all instances, but on the next levels, certain tests can be eliminated (the number of instances is reduced in each step), and only in the affected subtree do all the tests need to be performed (but on a reduced number of instances).

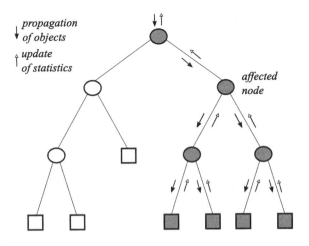

Fig. 8.6 A partial propagation of the training data in a decision tree after node modification

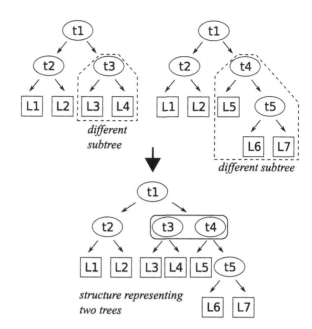

Fig. 8.7 Two trees with a common upper part (top), and these trees are represented in one structure (bottom)

Nodes on the lower parts of the tree are mutated with a higher probability, which favors the proposed optimization mechanism. In the case of classification trees, it is implemented [14] in such a way that the CPU sends only information about the affected subtree and the route from the root to this subtree to the GPU. Only tests on the path from the root node to the affected part of the tree (and inside this part) are needed to adjust the fitness of the original tree to reflect the current fitness.

The second observation is linked to similarities among the trees in the current population. In every selection, better individuals survive with a higher probability, and as a result, additional clones of existing trees appear in the population. Despite the application of genetic operators, the number of trees with common upper parts (i.e., the root and first split(s)) could be significant and may gradually increase. This natural process is especially visible when the evolution converges. In Fig. 8.7, an example of a situation where two trees share the test in the root node and the left subtree is shown. If the two trees' representation (structure) is implemented, it is possible to avoid repeated propagation of the learning instances in these common parts. As a cost, a little more complicated solution is necessary for the node where the trees differ. In Fig. 8.8, two possible situations in such a node are presented. Two separate tests are needed, or a test in one tree and a leaf in the second one (or vice-versa).

It should be noted that in the upper parts of the tree, the tests are evaluated on the important parts of the learning dataset (the whole dataset in the root node). This has a particularly large impact on the calculation time when large-scale data are analyzed.

The algorithm to identify tree pairs that can be combined into a common representation is rather straightforward. First, two trees need to have the same test in the root

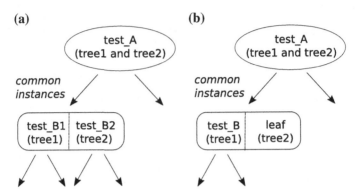

Fig. 8.8 Two situations that illustrate how two trees can differ in the internal node: **a** two different tests and **b** a test in one tree, with a leaf in the second tree

nodes. Then, we move down simultaneously on two trees, and every node difference (as discussed in Fig. 8.8) requires setting a kind of a forking node and finishing this path. If the decision trees are identical, there will be no forking nodes.

For a given population of trees, the pairs of similar trees can be easily selected based on the identity of the root test. It allows for creating a set of pairs that can be sent to the GPU as the two-tree representation for processing. The rest of the individuals can be served by the GPU in a standard way.

Here, the idea of a common representation of two similar trees can be easily generalized to a many-trees representation, but proper implementation is much more challenging.

8.2.2 Multi-GPU Implementation

A single GPU implementation, even though it has many advantages, has certain limitations, as demonstrated in the experimental results. The most important limitation is caused by the available memory constraints. If the size of an analyzed dataset is above tens/hundreds of millions of instances (depending on the hardware), the dataset cannot be loaded into the GPU's memory. Presently, more and more opportunities are emerging where a computational station can be equipped with more than one GPU. Such expandable multi-GPU systems offer a gradual and flexible mechanism to overcome the inherent limits of the simplest GPU-based solutions.

There are two natural approaches to the multi-GPU implementation of decision tree induction. In the first solution, each GPU processes a distinct subset of individuals from a population. If the number of GPUs is low (up to eight), at least a few individuals can be allocated to each card, and they can be processed one by one in the GPU. This approach seems to be easy to realize because it does not change processing on the GPU side, and the master process only needs to assign consecutive individuals to different GPUs. Moreover, if the number of individuals is divided by the number of cards without a remainder, the load imbalance most likely be mitigated if the tree sizes are similar in the population. Obviously, this approach should result in substantial speedup, but bigger datasets (which cannot be loaded into the GPU memory) are not able to be processed.

In the second approach, one individual is processed by all the cards in parallel and the individuals are processed one by one. A separate, equal-sized part of a training dataset is associated with each card. The master process broadcasts a tree to all the GPUs, and each card runs the same kernels as before on its own subset. The processed tree copies with all the derived statistics are sent back to the master, where a reduction is performed. Because all the tree copies have the same shape, the reduction should be short and simple. Certain small time loses can appear because of a load unbalancing, but they should be negligible because the subsets should be equal and pre-shuffled. Such an approach can be treated as an extension of the solution realized for a single GPU, where the idea of a training dataset partitioning (on blocks and threads) is just moved to a higher level (on GPUs). The main advantage of this approach is the possibility of processing much larger datasets because a dataset has to fit into the sum of the memories of all cards. It scales very effectively to a certain extent,[8] because additional cards reduce a dataset fraction processed by one card and enlarge the dataset size that can be mined. For datasets that can be loaded into the memory of a single card, the approach should also offer a real speedup, one maybe only slightly lower than the first solution.

From an implementation point of view, the second approach is relatively simple. Only small extensions of the single GPU code are necessary. Before the actual evolution, instead of transferring the full dataset, the dataset is divided into equal partitions, and each partition is sent to the corresponding GPU. During the evolution when the fitness of a tree is calculated, the tree is just broadcasted to all available GPU cards, and all GPUs work independently, exactly in the same way as in the single GPU case. For each GPU, one OpenMP thread is created. Only a reduction of the partial results from the GPUs needs to be additionally implemented, but it can be effectively combined with the already implemented tree decoding. On the algorithmic level, the only difference between the single- and multi-GPU implementations deals with the dipole searching mechanism because in the multi-GPU version, the dipoles are drawn only from instances belonging to a dataset partition associated with a given card. It seems that such a subtle difference is completely negligible from an evolutionary induction point of view.

8.2.3 Experimental Results

The adequacy of the GPU-based boosting of evolutionary decision tree induction can be best presented in an experimental way. First, in two sets of trials, the possible

---

[8]The number of GPUs that can be installed in one station is usually explicitly limited.

Table 8.2 Processing and memory resources of the NVIDIA GPU cards used in the experiments

GPU card	Engine		Memory	
	No. cores	Clock rate (MHz)	Size (GB)	Bandwidth (GB/s)
GeForce GTX 780	2 304	863	3	288.4
Titan X	3 072	1 000	12	336.5
Tesla P100	3 584	1 329	12	549.0

speedup for the classification trees' and model trees' generation on various GPU cards are measured. The time- and memory-related limits of the accelerated method are also studied. Finally, the multi-GPU approach is verified.

For single GPU experiments, three NVIDIA GPU cards are tested: GeForce GTX 780, GeForce GTX Titan X, and Tesla P100. Each of them is based on a different architecture: Kepler, Maxwell, and Pascal, respectively. The first two cards are the consumer line GeForce GPUs, while the third one is the professional-level GPU accelerator. At the time of writing this book, this card costs about $ 5000 (almost five times more than the second GPU and ten times than the first one). Table 8.2 presents more specification details.

The first two GPU cards are installed alternately in a workstation equipped with a processor Intel Xeon E5-2620 v3 (15 MB Cache, 2.40 GHz) and 96 GB RAM. The third one is installed in a multi-GPU server equipped with two processors of the Intel Xeon E5-2620 v4 (20 MB Cache, 2.10 GHz) and 256 GB RAM. In both machines, Ubuntu Linux 16.04.02 LTS (64-bit version) is used as the operating system. The GPU-based parallelization part is implemented in CUDA-C and compiled by nvcc CUDA 8.0 [20] (single-precision arithmetic is applied). Sequential versions of the algorithms, as well as OpenMP-based parallelizations, are run on the second machine because its CPU resources are stronger. Each Intel Xeon E5-2620 CPU is an eight-core processor.

Accelerating Classification Tree Induction

Let us first concentrate on boosting the evolutionary induction of the classification trees, where the proposed parallel implementation is rather straightforward. In Fig. 8.9, the obtained speedup for the *chess3x3* problem using three GPU cards is presented. The number of instances in the analyzed datasets grows from 1 000 000 (1 M) to 100 000 000 (100 M), whereas the number of features is fixed. The optimal decision tree for each dataset looks exactly the same (eight internal nodes and nine leaves) because all randomly generated instances are analytically assigned to follow the original two-class pattern. The global inducer finds the expected tree with no problems for all dataset variants.

Fig. 8.9 Speedup evaluation for various GPUs in classification tree induction. *Chess3x3* dataset variants are analyzed, with varying sizes from 1 to 100 M instances

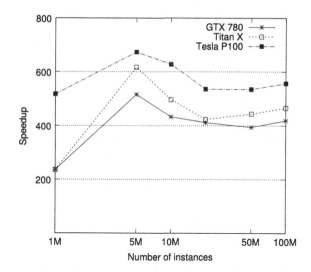

Table 8.3 Obtained speedup (classification tree induction) for the real-life datasets for different GPUs

Dataset	Speedup on different hardware			
	2×Xeon CPU	GTX 780	Titan X	Tesla P100
Suzy	7.0	574	645	807
Higgs	5.2	578	612	683

The obtained results are spectacular, especially if compared with the acceleration observed for the computing clusters. The speedup depends on the cards, but for all GPUs, it is around a few hundreds. The best-measured speedup is almost 700. The speedup profiles are rather similar for all the cards. The 1 M dataset does not seem big enough for the most efficient parallelization, especially in the case of the GTX and Titan X cards. The peak is observed for the 5 M dataset, and then, the obtained speedup slightly decreases, but it stays above 400, which is a very competitive result.

Next, two large real-life datasets from the UCI Machine Learning Repository [13] are analyzed: *Suzy* (5M instances, eighteen features, two classes) and *Higgs* (11 M instances, twenty-eight features, also two classes). The obtained speedup for these datasets is presented in Table 8.3. Additionally, because the cards were installed and tested on multi-core machines, the speedup obtained without GPU acceleration but rather with OpenMP is provided. We can clearly see that even when using the cheapest GPU, the measured speedup is two orders of magnitude higher.

Let us discuss the absolute induction times to better show the difference introduced by GPU-based acceleration. The sequential version of the GDT system on Intel Xeon CPU needs about 160 000 s (two days) for *Suzy* and 350 000 s (four days) for *Higgs*, whereas the shortest time obtained for a GPU-based solution is only 197 s (3.5 min) and 510 s (8.5 min), respectively. Taking into account the dataset size and the involved hardware price, the obtained execution time reduction is very substantial.

Fig. 8.10 Time-sharing information (mean time as a percentage) of GPU-accelerated classification tree induction on the Tesla P100. The *chess3x3* dataset variants are analyzed with varying sizes ranging from 1 to 100 M instances

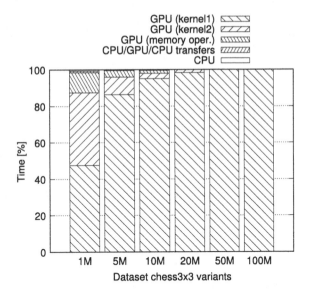

To understand how such a huge acceleration is possible, we decide to investigate time-sharing information in detail. In Fig. 8.10, the execution time is divided into five blocks corresponding to two kernels, memory operation on the GPU, data transfers between the CPU and GPU, and instructions running only on the CPU. The same series of *chess3x3* datasets with an increasing number of instances is analyzed, and for each dataset, the time fractions devoted to these blocks are depicted. Certain parts of the algorithm are dependent on the dataset size, whereas other parts are related more to tree sizes. Moreover, the non-parallelizable parts of the code are short. For the smallest dataset, the time spent on two kernels is roughly equal, and the data transfer time is also noticeable. With increasing dataset size, the first kernel time becomes dominant, and from 50 M instances, the other operation time can be completely ignored. During the execution of the kernel, a few thousand GPU cores can work in parallel, which explains the good resulting speedup.

The speedup offered by the presented parallel implementation can be further improved. The first of the aforementioned improvements is experimentally validated, and the relative speedup increase is presented in Fig. 8.11 for the various GPU cards.

The partial tree recalculation brings noticeable savings for all cards. For the smallest dataset, it increases the speedup by only 5.5% on the GTX 780 card, but typically, it is about twenty percent, and for the biggest datasets, it can be even more that thirty percent.

Furthermore, the influence of the number of features on the speedup is also investigated. Once again, the *chess3x3* problem is studied, and to each dataset, random features not related to the classes are added (from eight up to 998 additional features). In Fig. 8.12, the obtained speedup for three moderate dataset sizes (number of instances: 100, 500 K, and 1 M) and the increasing number of features on the Tesla P100 card are presented. The partial tree recalculation is switched on.

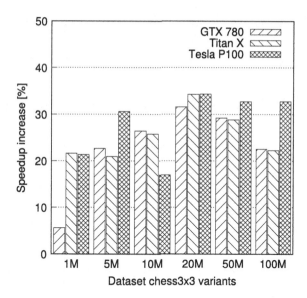

Fig. 8.11 Relative speedup increase for various GPUs in classification tree induction from processing only modified part of the trees. The *chess3x3* dataset variants are analyzed with varying sizes ranging from 1 to 100 M instances

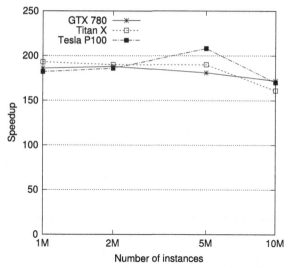

Fig. 8.12 Speedup evaluation for the Tesla P100 in classification tree induction. The *chess3x3* dataset variants are analyzed with a varying number of features (from 2 to 1000)

Here, the observed speedup grows with the number of features. Each additional feature in a dataset increases the memory requirements, and they are much more important for CPU-based induction. In the case of the GPU-accelerated version, the iteration execution time becomes just a little longer, and this results in a visible increase of the speedup. However, a high number of nonrelevant and noisy features obviously slows down the evolutionary induction, and more iterations may be necessary to obtain a competitive classifier.

Table 8.4 The maximum size of the *chess3x3* dataset variant that can be completely processed in the given period of time by the GPU-accelerated GDT system. In addition, the results for the sequential CPU version and the OpenMP parallelization are included

GPU card/period	1 min	1 h
GTX 780	530 000	57 000 000
Titan X	1 124 000	76 000 000
Tesla P100	966 000	84 000 000
Sequential CPU	1 200	275 000
2×Xeon CPU (16 cores)	61 000	1 150 000

Table 8.5 The maximum size of the *chess3x3* dataset that can be completely processed by the given GPU card. The processing time is also included

GPU card	Dataset size	Time (h)
GTX 780	256 000 000	5
Titan X	1 033 000 000	21
Tesla P100	1 033 000 000	17

Finally, let us check where the limits of the GPU-accelerated evolutionary induction in terms of the dataset size that can be processed in the given period of time are [21]. The results are obtained with the partial tree recalculation. Two periods are considered: one minute and one hour. In Table 8.4, the maximum size of the *chess3x3* dataset, which can be completely processed, is presented.

It is revealed here that in one minute, the dataset composed of about a half million of instances can be analyzed with the weakest card, and a dataset that is twice as large can be processed by stronger cards. It is substantial progress if compared with 1200 instances, which can be mined by the sequential version. For the one-hour period, the limit grows to more than fifty million instances.

It is worth remembering that the dataset size restrictions are also related to the size of the available memory on the GPU cards. In Table 8.5, the maximum sizes of the *chess3x3* dataset that can be mined with the GPU cards are presented.

Here, a dataset with more than one billion instances can be processed by the cards with the biggest memory, and it takes about twenty hours. For less-equipped cards, the maximum dataset size is four times smaller, but still, the numbers are impressive.

Accelerating Model Tree Induction

Model tree induction is naturally much more computationally complex because it incudes searching for linear models in leaves. Moreover, the estimation of the prediction error is more difficult because it requires the calculation of residuals for every training instance using the corresponding model. These important differences in the sequential algorithm and the parallel implementation entail clearly visible changes

Fig. 8.13 Speedup evaluation for various GPUs in a model tree induction. The *Armchair* dataset variants are analyzed with varying sizes ranging from 1 to 10 M instances

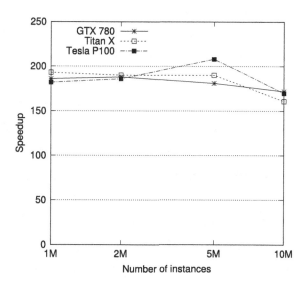

in the observed speedup. Below, some selected results are presented to show the differences.

In Fig. 8.13, the obtained speedup on the three considered GPU cards are presented. The *armchair* regression problem [22] with an increasing number of instances (from 1 M instances to 10 M) is studied.

It can be easily observed that the speedup is not as huge as in the case of classification tree induction, but still, it is very competitive. Interestingly, the differences between the cards are less visible, which favors the cheapest card.

Much like for the classification tree induction, two large real-life datasets [13] are analysed: *Year* (515 345 instances, ninety real-valued features) and once again *Suzy* (5 M instances, seventeen features). Because of the lack of publicly available large-scale regression datasets, *Suzy* is slightly transformed in such a way that the value of the last feature is predicted, and the original class is eliminated. This modified dataset version is denoted as *Suzy'*. The only purpose of this operation is to investigate the algorithm's speedup, not the prediction performance. The obtained speedup for these two datasets is presented in Table 8.6.

The obtained speedup is visibly better compared with the previous experiment on the artificial datasets, but the number of features in the analyzed datasets has increased (from two up to ninety features), and this could explain the vast improvement. The high number of features seems to be the most important for calculating the models in leaves. Again, all three cards offer very similar performance, which is rather surprising.

Finally, to better understand the specificity of parallelization of model tree induction, a time-sharing graph has been prepared for the Tesla P100 card (Fig. 8.14). Four

Table 8.6 Obtained speedup (model tree induction) for real-life datasets when using different GPUs

Dataset	Speedup on different hardware			
	2×Xeon CPU	GTX 780	GTX Titan X	Tesla P100
Year	4.9	346	351	367
Suzy'	4.2	720	736	742

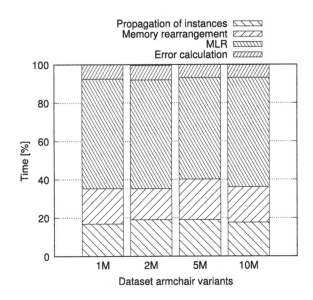

Fig. 8.14 Time-sharing information (mean time as a percentage) of the GPU-accelerated model tree induction on the Tesla P100. The *Armchair* dataset variants are analyzed with varying sizes ranging from 1 to 10 M instances

main groups of activities are distinguished: in the first, the training instance locations in a tree are determined, the second encompasses the necessary reorganizations of the data structures and memory for launching a linear regression fitting, the third is the actual fitting based on the optimized library procedures, and in the last group, the residuals are calculated and reduced.

The most obvious observation is that all the profiles are very similar, and the size of the dataset seems not to affect the time sharing. This pattern is completely different from the one observed in the case of classification tree induction, where the induction time is dominated by one activity for larger datasets. Here, the time sharing is much more uniform. All phases are dependent on the dataset size, and it explains why the obtained speedup is rather stable.

More results on the GPU-based acceleration of regression tree induction can be found in [15], where, among others, various blocks and threads configurations are studied.

Fig. 8.15 Speedup evaluation for the multi-GPU implementation of classification tree induction. The *chess3x3* dataset variants are analyzed with varying sizes ranging from 1 to 1000 M instances

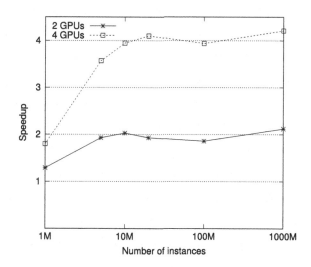

Multi-GPU Acceleration Results

The second of the aforementioned multi-GPU approaches is realised (dataset decomposition), and its performance is evaluated on a server equipped with two processors of the Intel Xeon E5-2620 v4 (20 MB Cache), 256 GB RAM, and four NVIDIA Tesla P100 GPU cards. Each CPU contains eight physical cores running at 2.10 GHz. Each GPU chip has 3584 CUDA cores and 12 GB of memory. The same operating system and compiler are used as for single-GPU experiments. The multi-GPU implementation is confronted with the single-GPU implementation on previously analysed variants of the *chess3x3* dataset with varying sizes. The obtained speedup with two and four cards (over one GPU) is presented in Fig. 8.15.

For bigger datasets (10 M instances and more), the obtained speedup is almost perfect, and the slight fluctuations can be explained by a stochastic character of the algorithm. For the smallest dataset variants, the observed performance is not so satisfying because the overhead is too important. The smallest datasets may not be able to saturate the large number of cores in the multiple GPUs.

8.3 Spark-Based Solution

GPU-related adaptations of the global induction algorithm lead to the isolation of the fitness function calculation because it is the most time-consuming operation. Moreover, the training dataset can be directly accessed only when the fitness is estimated, and its distribution in each tree is recalculated on-demand with the GPU kernels. This indicates that the GPU card-based mechanism can be replaced with any other parallel/distributed system, like Spark, which has emerged recently as a

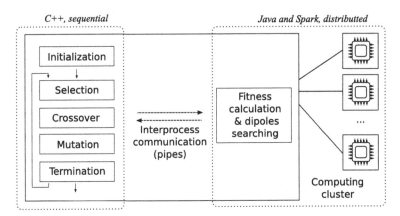

Fig. 8.16 Architecture of the Spark-accelerated implementation of evolutionary decision tree induction [23]

new tool for data-intensive computation on computing clusters. It can be especially attractive if the GPU memory constraints are exceeded on the available hardware. For example, every new feature in the processed dataset reduces the maximal number of instances that can be mined. It can be expected that the processing time will be longer, but the scaling properties of the Spark-based approach (just by adding new nodes to the computing cluster) are significantly larger.

On the conceptual level, the idea of incorporating Spark for fitness processing seems to be straightforward, but the actual realization is not so easy. First, the original GDT system is a native C++ application, whereas Spark (written in Scala) uses a multi-process architecture and runs its processes on the Java Virtual Machine. Despite this discrepancy, a smart solution can be developed based on efficient interprocess communication mechanisms. In Fig. 8.16, the proposed heterogeneous architecture [23] is presented. A new Java-based module needs to be introduced, which reimplements only a small fraction of the procedures dealing with the fitness calculation and the training dataset access. The main Spark process is known as a driver, and it dispatches the work to multiple processes (workers) distributed over the available nodes of the computing cluster. Hence, the new module plays the role of the driver and enables a distributed propagation of instances through a tree and the dipoles searching in the same moment. The rest of the evolutionary algorithm is untouched and it is realized in a separate process (C++). The Spark driver and the GDT application are launched on the same machine, and they communicate and transfer the data through the named pipes.

The key issue for a Spark-based algorithm is the proper definition and utilization of an RDD, which represents the training dataset. All the training instances are loaded into the RDD and are assigned to groups of an equal size. Depending on the size of the training dataset, two methods can be applied:

- `SparkContext::parallelize()` (from the Spark API) loads a local file and distributes the RDD on the cluster. The method is appropriate for smaller datasets;
- `SparkContext::textFile()` enables loading from HDFS and is more convenient for larger files.

Regardless of the method used, the input is processed line by line, and each file line is parsed into one instance. It may seem that each instance should correspond to an object in RDD, but such a simple association is not optimal from a memory usage standpoint, especially for large-scale data. Instead, the instances are loaded into the objects representing packages of instances (1 000 instances in a package by default). This results in a significantly reduced number of objects in the RDD, which in turn minimizes the overhead of the RDD data structures. Each package is randomly assigned to a group with an unique identifier (`groupID`), and this creates a key-value RDD of (`groupID, package`) elements. The number of groups should correspond to the number of data partitions that will be distributed and processed. In the simplest scenario, the number of partitions is equal to the number of available workers. It can also be related to the number of available cores in the cluster. Then, `RDD::groupByKey()` is evoked, and it triggers a global dataset repartition. As a result, the training dataset is uniformly distributed over the entire cluster and can be cached in the memory of the computing nodes. Because the evolutionary algorithm is highly iterative, the proper distribution of the dataset chunks in the cluster memory is one of the most crucial implementation aspects. It should be emphasized that the repartition operation is performed only once, and the cost of this is negligible especially, when we take into account the effects.

When the data are distributed, the actual processing of the requests from the C++ GDT process can be realized. Every request received by the Spark driver deals with a single decision tree. This requires the passing of all training instances through the received tree to obtain their locations in leaves, which allows for calculating the class distributions. In Fig. 8.17, applied parallel processing is depicted, here consisting of a typical pair of `map-reduce` operations evoked on the grouped RDD. First, each group emits a locally processed copy of the tree (`map(group)` → `tree`). In the second step, the local trees are reduced into the globally processed tree (`reduce(tree, tree)` → `tree`). During the reduction, the class distributions in every leaf are simply merged. With each local copy of the tree, a set of dipoles that may be used in the next iteration by genetic operators is maintained. The dipoles are reduced implicitly by selecting them from one of the trees. Finally, after additional inner-tree processing (certain elements' propagation from leaves up to the root), the resulting tree can be sent back to the C++ GDT process.

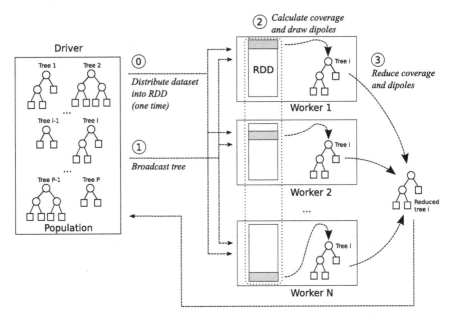

Fig. 8.17 Distributed calculation of the fitness for a single decision tree. The possible dipoles for genetic operators in the next iteration are also randomly chosen

8.3.1 Experimental Results

The same artificial and real-life datasets are analyzed as in the previous section, where the classification tree induction is accelerated with the GPU cards. And similarly, all the presented results are obtained with a default set of parameter values from the sequential version of the GDT system. Only the induction times and speedup are provided because the proposed solution does not really affect the resulting classifiers.

A cluster of eighteen SMP workstations (with quad-core Intel Xeon E3-1270 3.4 GHz CPU, 16 GB RAM) connected by a Gigabit Ethernet network and running Ubuntu 16.04 (Linux 4.4) is used for the experimental works. One node is dedicated to Spark Master and HDFS NameNode. The second node is running Spark Driver and GDT C++ processes, and the remaining sixteen worker nodes are used by Spark executors. The cluster is managed with Apache Hadoop 2.7.3, and Apache Spark 2.2.0 [24] is deployed in a stand-alone mode with a single executor on each worker node.

In Fig. 8.18, the speedup results obtained with the Spark-based solution for classification tree induction are presented. Three different numbers of workers are analyzed. Each Spark worker corresponds to one SMP processor.

Generally, the observed speedup is only acceptable because after knowing the performance of the GPU-based approaches, the expectations are simply very high. If we recall that each processor is equipped with four cores, the obtained speedup

Fig. 8.18 Speedup evaluation of the Spark-based solution in a classification tree induction. The *Chess3x3* dataset variants are analyzed with varying sizes ranging from 1 to 100 M instances

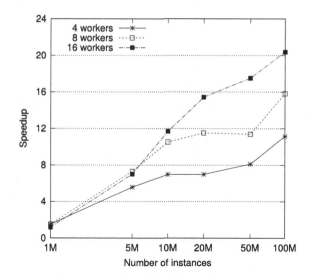

Table 8.7 Obtained speedup of the Spark-based solution in a classification tree induction for two real-life datasets with a different number of workers

Dataset/speedup	Number of workers		
	4	8	16
Suzy	5.80	8.16	8.03
Higgs	6.15	10.22	14.23

looks worse. However, Spark is not intended to be extremely fast but rather a reliable solution for Big Data processing. The dataset size strongly influences the possible acceleration. For the smallest dataset (1 M of instances), the Spark-based boosting does not make sense, but with an increase in the size, the observed speedup grows dynamically. Also, adding more workers is profitable only for large datasets.

The Spark-based solution is also verified on two real-life datasets. In Table 8.7, the results with the same number of workers are presented. Concerning four and eight workers, the obtained speedup is even greater than the number of processors involved, but each processor has four cores. Regarding the instance where there are sixteen workers, the speedup obtained for the *Higgs* dataset is proper, but for the smaller dataset, it seems that a slightly increased overhead cancels out the speedup gain [23].

It could be interesting to confront the Spark-based global decision tree induction with a distributed version of the classical top-down induction algorithm available in Spark's MLlib library [25]. To have at least a rough comparison, the basic decision tree inducer is applied (with default settings) to mine these two real-life datasets. The obtained accuracies are practically the same, but the GDT generated substantially smaller trees (from five to seven times smaller number of nodes). On the other hand, the induction time using all the cluster's machines is about thirty minutes for the GDT and only one to two minutes for MLlib, but this is not surprising.

Table 8.8 The maximum size of the *chess3x3* that processed by the Spark-based implementation for the different numbers of workers in the given period of time

Workers/period	1 h	1 day
4	13 000 000	500 000 000
8	20 000 000	1 250 000 000
16	35 000 000	2 500 000 000

Table 8.9 The maximum size of the *chess3x3* dataset that can be completely in-memory processed with the Spark-based solution and the given number of workers. The processing time is also included

Workers	Dataset size	Time (h)
4	900 000 000	33
8	1 850 000 000	35.5
16	3 900 000 000	38

As the last element, let us verify the limits of the Spark-based evolutionary induction in terms of dataset size, which can be processed in the given period of time [21]. This time, the considered periods have to be obviously longer than in the case of the GPU: one hour and twenty-four hours. It is worth mentioning that one minute is just enough to actually initiate the global induction on the Spark engine. In Table 8.8 the maximum size of the *chess3x3* dataset that can be completely processed is presented.

Here, the Spark-based solution offers an impressive number of instances that can be processed in an hour, and the size of the datasets nicely scales with the number of workers involved. The size of the dataset that can be mined (in the same period) is only a few times smaller compared with the GPU-based version. Also, if the computation time is not so important, the offered induction possibilities grow impressively.

In the Spark-based approach, the dataset size restrictions are closely related to the number of workers and their memories. In Table 8.9, the maximum sizes of the *chess3x3* dataset that can be mined (in-memory mode) with a varying number of workers are presented. The larger datasets can also be processed, but a significant drop in efficiency will be observed because of the necessary swapping (memory to disk) activities.

It can be seen that a dataset with about four billion instances can be processed by the tested computing cluster, but the computation time is really long. This is without doubt a convincing example of Big Data mining. A raw text file size corresponding to this dataset is about 78 GB, and in memory-cached RDD requires 96 GB. Interestingly, adding more worker nodes to the cluster linearly increases the maximal size of the dataset and only slightly extends the induction time.

As a final remark, the proposed architecture of the Spark-based solution with a clear separation of the evolution and fitness evaluation offers many chances for further performance improvement. For example, the natural similarity of the processed individuals, which results from selection mechanisms and limited modifications dur-

Fig. 8.19 Relative execution time decrease for various numbers of Spark workers in classification tree induction because of processing only modified parts of the trees. The *chess3x3* dataset variants are analyzed with varying sizes ranging from 1 to 100 M instances

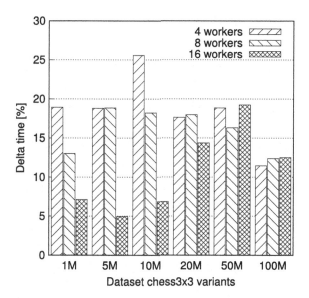

ing genetic differentiation, leads to situations where the same or very similar trees need to be reexamined. On the Spark side, it can be profitable to maintain a buffer of previously processed trees. Before processing a newly received tree, the buffer can be efficiently searched, and if a similar tree is found, only partial calculations are necessary. This idea can be further extended and the concept of multi-tree representation can be easily incorporated. As a kind of an incentive, let us present the results of the same improvement as presented for the GPU-based implementation of classification tree induction, where the request to process only a modified part of the tree can be accepted by the Spark driver. In Fig. 8.19, the obtained difference in execution time for various numbers of workers is presented on the *chess3x3* datasets.

Here, even such a limited improvement can give noticeable time gains (about fifteen percent). In this simple proposition, only about fifty percent of the requests sent to the Spark driver are for partial tree calculations, whereas the rest are calculated in the standard way (e.g., a mutation of a test in a root node excludes the partial processing). This suggests that there are still many opportunities for improvement.

References

1. Alba E (2005) Parallel metaheuristics: a new class of algorithms. Wiley, New York
2. Alba E, Tomassini M (2002) IEEE Trans Evol Comput 6(5):443–462
3. Gong Y, Chen W, Zhan Z, Zhang J, Li Y, Zhang Q, Li J (2015) Appl Soft Comput 34:286–300
4. Kretowski M (2008) Obliczenia ewolucyjne w eksploracji danych. Globalna indukcja drzew decyzyjnych, Wydawnictwo Politechniki Bialostockiej
5. Alba E, Luque G, Nesmachnow S (2013) Int T Oper Res 20:1–48

6. Tsutsui S, Collet P (2013) Massively parallel evolutionary computation on GPGPUs. Springer, Berlin
7. Kretowski M, Grzes M (2007) Int J Data Wareh Min 3(4):68–82
8. Kalles D, Papagelis A (2010) Soft Comput 14(9):973–993
9. Kretowski M, Popczynski P (2008) Global induction of decision trees: from parallel implementation to distributed evolution. In: Proceedings of ICAISC'08. Lecture notes in artificial intelligence, vol 5097, pp 426–437
10. Czajkowski M, Jurczuk K, Kretowski M (2015) A parallel approach for evolutionary induced decision trees. MPI+OpenMP implementation. In: Proceedings of ICAISC'15. Lecture notes in artificial intelligence, vol 9119, pp 340–349
11. Czajkowski M, Jurczuk K, Kretowski M (2016) Hybrid parallelization of evolutionary model tree induction. In: Proceedings of ICAISC'16. Lecture notes in artificial intelligence, vol 9692, pp 370–379
12. Kretowski M (2008) A memetic algorithm for global induction of decision trees. In: Proceedings of SOFSEM'08. Lecture notes in computer science, vol 4910, pp 531–540
13. Dua D, Karra Taniskidou E (2017) UCI machine learning repository. Irvine, CA: University of California, School of Information and Computer Science. http://archive.ics.uci.edu/ml
14. Jurczuk K, Czajkowski M, Kretowski M (2017) Soft Comput 21:7363–79
15. Jurczuk K, Czajkowski M, Kretowski M (2017) GPU-accelerated evolutionary induction of regression trees, In: Proceedings of TPNC'17. Lecture notes in computer science, vol 10687, pp 87–99
16. Singh D, Joshi I, Choudhary J (2018) Int J Parallel Program 46(6):1017–1034
17. Merrill D (2018) CUB v1.8.0 A library of warp-wide, block-wide, and device-wide GPU parallel primitives, NVIDIA Research. http://nvlabs.github.io/cub/
18. NVIDIA (2018) cuBLAS, NVIDIA developer zone, CUDA toolkit documentation. https://docs.nvidia.com/cuda/cublas/
19. NVIDIA (2018) cuSOLVER, NVIDIA developer zone, CUDA toolkit documentation. https://docs.nvidia.com/cuda/cusolver/
20. NVIDIA (2018) CUDA C programming guide. http://docs.nvidia.com/cuda/pdf/CUDA_C_Programming_Guide.pdf
21. Jurczuk K, Reska D, Kretowski M (2018) What are the limits of evolutionary induction of decision trees? In: Proceedings of PPSN XV. Lecture notes in computer science, vol 11102, pp 461–473
22. Czajkowski M, Kretowski M (2014) Inform Sci 288:153–173
23. Reska D, Jurczuk K, Kretowski M (2018) Evolutionary induction of classification trees on Spark. In: Proceedings of ICAISC'18. Lecture notes in artificial intelligence, vol 10841, pp 514–523
24. The Apache Software Foundation (2019) Apache spark - lightning-fast cluster computing. http://spark.apache.org/
25. Meng X et al (2016) J Mach Learn Res 17(1):1235–1241

Concluding Remarks

Evolutionary algorithms are often appreciated as robust and flexible meta-heuristics; they are successfully applied to a very broad spectrum of optimization and search problems. If we understand the specificity of the analyzed data, we can typically use this knowledge to improve the algorithm by introducing specialized genetic operators and/or defining the precisely targeted fitness function. Even if the number of evolutionary algorithm parameters are high, they can be easily fine-tuned because the results of the algorithm are usually stable in a wide range of reasonable parameter values. On the other hand, evolutionary approaches are perceived as slow methods because they are stochastic and with a lot of randomness, and they process many potential solutions. This is the main argument against applying evolutionary computations in data mining systems. It is especially discouraging because solving Big Data problems is the ultimate goal for most modern, real-life systems.

In the book, I have tried to break apart this stereotype and to show that evolutionary data mining systems can be efficiently applied to large-scale data. I believe that at least for decision tree induction, this has been clearly realized. I demonstrated that the various types of a decision tree, starting from the simplest univariate classification trees through oblique or multi-test trees and up to mixed model trees for regression problems, can be globally induced. Undoubtedly, the evolutionary-generated decision trees have many advantages: they are accurate and simple, they can be easily understood and interpreted, and they can be problem-specific tailored. In the case of more and more common multi-objective problems, if an analyst has difficulties in comprehending the data and in deciding how to attack the problem, he or she can convert these doubts into a strength by applying the Pareto-based approach. This would obtain many potential solutions with varying trade-offs and being able to choose the most adequate one. But most of all, we managed to reveal that the evolutionary mining of huge datasets with millions (or even billions) of instances does not need any extraordinary resources. With the smart use of current GPUs, which offer simplified capabilities but still have extreme computational power, the speed of the global induction increases a great deal. If the size of the datasets makes the fastest in-memory computation impossible, the Spark-based implementation on commodity

© Springer Nature Switzerland AG 2019

M. Kretowski, *Evolutionary Decision Trees in Large-Scale Data Mining*,

Studies in Big Data 59, https://doi.org/10.1007/978-3-030-21851-5

computer clusters comes in play and enables us to break down the next barriers. The induction becomes obviously slower, but the fault tolerance and scalability potential are impressive. It should be underlined that both these boosting technologies have very attractive cost-to-power ratios, and they are accessible, even for small companies and organizations.

Recently, a lot of attention has been given to the spectacular achievements of deep learning in computer vision applications [1]. Artificial intelligence systems enhanced by GPU-based machines can automatically recognize a type of tumoral change on tomographic images or autonomously drive a car. It seems that it is not significant methodological advances but just substantial progress in computational resources (both on the software and hardware level) that has really moved the approach forward. I perceive some analogies between a situation of deep learning and evolutionary data mining, and I think that similar progress is possible thanks to the effective use of available computational power.

Future Directions

So far, the acceleration of evolutionary decision tree induction has been focussed on increasing the number of instances that can be mined. Now, it is clear that from a computational point of view, really huge datasets can be processed, and this dimension can be mastered with dataset division techniques. Gene expression data signals the rising impact of the second dimension and the influence of the number of features on the reliability and robustness of evolutionary mining, and this needs to be further studied. Generally, it can be expected that the evolution will be less efficient with an increase in the number of features. The algorithms may require more individuals and/or more iterations, which of course increases the demand for computational power. Once again, new parallel and distributed solutions should be investigated.

The complexity of the predictive problem can be measured by the size of the model, which offers satisfactory predictions. It could be very interesting to investigate how an evolutionary approach will work with more complex problems that need trees with hundreds or thousands of nodes.

Global induction clearly leads to better predictors: the resulting trees are usually smaller, more stable, and often more accurate. The separation of the actual evolutionary algorithm and the fast fitness calculation in parallel systems offers a good framework for mining other predictive structures. A similar approach can be developed to evolve multi-layered neural networks, Bayesian networks, or sets of decision rules.

In the existing realizations of evolutionary decision tree induction, all the tests, regardless of the number of features involved, are crisp. On the other hand, soft or fuzzy splits can be successfully introduced into decision trees. By the integration of the partial routing of an instance, the problem of missing values can be solved.

In many applications of a data mining approach to Big Data, the data source can have a very dynamic nature. It may encompass classical video streaming, online financial time series, or the propagation of (fake) news in social networks. If the changes are fast, it may not be possible to relaunch the learning after each major change, so it may be profitable to develop methods that can adapt the existing model to new data characteristics. It seems that evolutionary induction can be involved this type of instant learning.

Reference

1. LeCun Y, Bengio Y, Hinton G (2015) Nature 521:436–444

Index

© Springer Nature Switzerland AG 2019
M. Kretowski, *Evolutionary Decision Trees in Large-Scale Data Mining*,
Studies in Big Data 59, https://doi.org/10.1007/978-3-030-21851-5

Printed in the United States
By Bookmasters